HARNESS THE JUICE

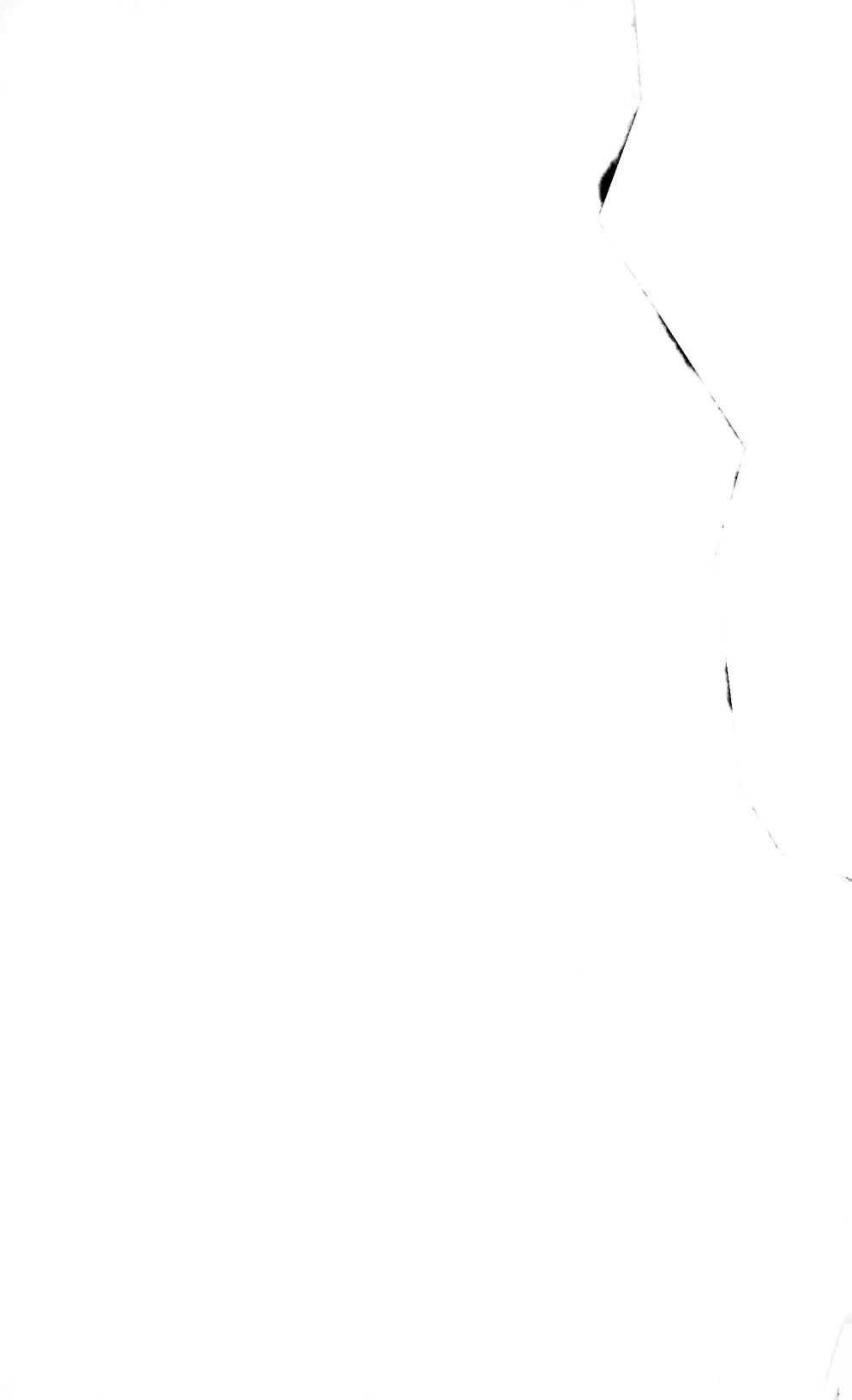

Harness
the Juice

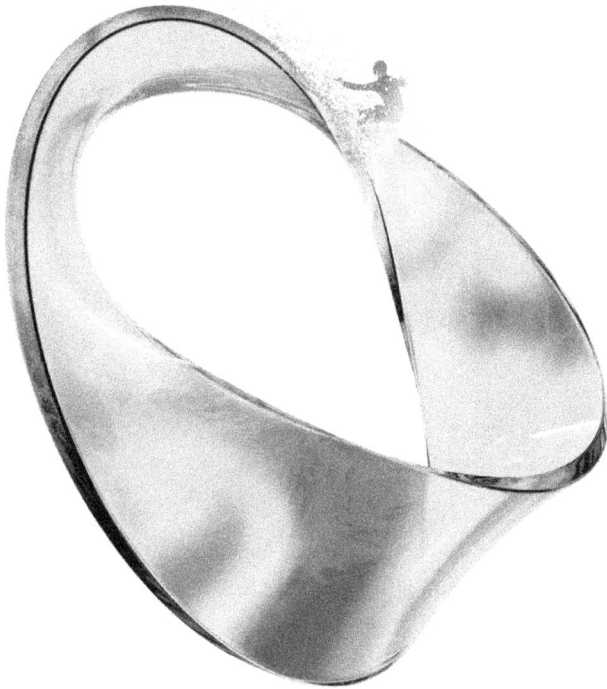

How to Surf the Impending
Tsunami of Technological
Innovation without Wiping Out

Aaron Alfini

LIONCREST
PUBLISHING

HARNESS THE JUICE

How to Surf the Impending Tsunami of Technological Innovation without Wiping Out

FIRST EDITION

ISBN   978-1-5445-3982-9 *Hardcover*
      978-1-5445-3983-6 *Paperback*
      978-1-5445-3984-3 *Ebook*
      978-1-5445-3985-0 *Audiobook*

To my first Mentor, Marianne

Your guidance set me on the path to where I am today. I can't complain.

Contents

Introduction

Waves are funny things. The same waves can be both soothing *and* horrifying. It just depends on the circumstance. You could be lying on the beach, soaking up the sun and listening to the gentle lapping of the waves on the shore, lulling you into peace and contentment. On the other hand, you could be terrified, caught in a current, heading for a shoreline of rocks with water crashing into them, paralyzed in fear, knowing that the waves will bash you into a bloody mess. Your circumstances determine how you feel about the waves.

The most dangerous waves are the ones that you can't see. The ones that are hidden below the water's surface before they reach the shore, when it's too late to react. A tsunami is that kind of wave.

Technology is also a wave. Technically, a wave transfers energy—through a medium—from one location to another. In the case of technology, the medium is time, and the first beginning location is the start of human existence. The next location? Our future. The energy being transferred is the innovation that we humans put into technology.

The amount of energy a wave has depends on how much energy was put into it. With ocean waves, wind is typically that source of energy. So, the power of waves ebbs and flows based on the power of the wind pushing the water. The technology wave primarily works in the same way; energy in equals energy throughout. However, there is a significant difference. As humans have progressed, we have put, and continue to put, more and more innovative energy into the wave. The technology wave doesn't ebb and flow. It simply grows and compounds.

Right now, we're at an apex. We have put so much energy into the wave of technology that it's become a tsunami. I'm writing this in 2022; within the next few years, we'll see a wall of water coming at us with unbelievable speed and power. The problem is, most people don't even know the tsunami's coming. Right now, it's just starting to become visible, far out in the distance. But few see it; a swell is only visible if you know what you're looking for.

When the wave does reach shore, many business owners, managers, and companies will be caught off guard. They will be so suddenly swept away by technology and fierce competition, that they won't even have a chance to react. Companies like Kodak, Blockbuster, Toys R Us, and Borders are just a few examples that have met their fate. But those companies had plenty of time to respond. In the future, others will be washed away with lightning speed. You must learn to adopt technology quickly and effectively to surf this wave if you want to save (and grow) your business.

To successfully adopt technologies, companies and their leaders must understand what the technology tsunami is, and why acting *now* is so important. It's also helpful to understand how technology has affected companies in the recent past, and why and how technology's impact on companies is accelerating.

Armed with an understanding of what is happening technologically, how to react, and when to implement change, you'll have a solid foundation to ride the wave. Whether you're a manager, executive, business owner, or board member, with the right preparation, you can know how to address technology adoption within your organization.

Technology adoption isn't just about information technology (IT), and it doesn't only affect one department. It includes you having the right mindset and understanding the mental barriers to successful preparedness and innovation. Only then can you progress toward addressing the people, process, and hidden dangers in your company.

WHAT WE'LL LEARN TOGETHER

In this book, you will learn how to surf the tsunami. How to take your business into the future and not only survive but thrive. In surfer parlance, "juice" refers to the power of a wave. I want you to be able to harness the juice of the tsunami and use it to your competitive advantage. You will learn how to think about technology as more than a necessary evil and cost center, but as a potential source of revenue or revenue expansion. I will highlight the barriers to technology adoption that you might not even know exist. What's more, I'll give you options on how to address them. *You* will be ahead of the wave.

Throughout my career, I've seen general technology, and many specific innovations surge forward. I have seen the rise of the technology tsunami firsthand. Not to date myself, but I first learned to program on a Commodore 64 attached to a black and white TV. Those first experiences ignited a fire in me that has only grown. I got my first job when I was buying parts to build my first computer in high school. The owner said, "You seem to know what you are talking about. Would you like a job?" The rest is history.

Since then, I've been a trusted advisor driving companies to the cutting edge of technology: replacing mainframes with client-server technology, being a pioneer in virtualization, and performing as one of the foremost experts on enterprise cloud migrations. I consider myself blessed, having experienced so many different technologies in my career. I've worked in healthcare, retail, marketing, financial services, and with some of the greatest brands in the world, such as Amazon, Walgreens, Discovery Communications, Western Union, The London Underground, Equifax, Allscripts, and Splunk. I even interviewed at the CIA. Even though I didn't get in (thanks, Snowden), even stepping into the building for an interview is a feat few have accomplished.

I have seen many technologies come and go over the years. Some companies have risen to the occasion, and led. Others have been disrupted. Now, the *pace* has changed: there's an ever-increasing acceleration factor. While many companies have created or at least embraced this bright new future, I've grown increasingly concerned that many great companies have been surprised by it, failing to properly adopt technology and build an internal culture of innovation. And *that* is why I've written this book.

I want to help owners, leaders, and companies survive the impending tsunami. Not only that—I want you to ride it.

Today, a small- to medium-sized company could be out of business in a matter of months if they can't adapt. With the COVID-19 pandemic, we had a concentrated snapshot of technology's impact on survival:

Many companies, such as certain restaurants, adopted new technology to deliver their products to customers, and they survived. In some cases, they even developed new avenues for growth. Meanwhile, those who didn't adapt, perished.

I don't want to see your company fail. I want you to thrive. Not just for the owners' sakes, but for the people who work there, who need that paycheck to feed their families. There's no reason you need to fail, as long as you're adequately prepared. And you will be, if you keep reading.

This book is not a paint-by-numbers "do this, and all your problems will be gone" answer guide. Technology is vast, and the specifics often depend on industry, culture, competition, and more. Instead, this book paints a picture of what *good technology adoption* looks like, in a form that you can consume to address your company's specific needs.

Are you ready to survive the tsunami and amp up technology in your business? Turn the page, and join me in Chapter 1.

PART I

Read the Waves

CHAPTER 1

The Impending Tsunami

If you do nothing, you are going to drown. Not in a bathtub, or a pool, or even a lake. A tsunami is going to crush you and sweep you away. That is, if you don't take action. This tsunami isn't one made out of water, but rather one of technology and innovation, crushing your business and livelihood. It doesn't matter if you're a small business owner or a manager in a massive multinational. It's your duty to understand how the rapid growth of innovation will impact you, and your company.

Today is the day for action. The world is advancing technology at ever-increasing speeds. Every day, thousands of discoveries and innovations are taking place. Those discoveries lead to even more discoveries, and so on. It's an exponential curve. Right now, we are at a critical point on that curve—the point where technology will make huge impacts on organizations that fail to keep up. The time to prepare for the impending wall of water is now.

The good news is, you can do something about this wave. Like most things in life, you can use knowledge to get ahead of the wave. That is what this book will do for you—it will give you the gift of technology foresight. This foresight will create your technology surfboard. Not only will riding the technology wave prevent you from being put out on the street, but it will also allow you to reach new heights in your career or company as well.

Throughout my career, I've helped many companies drive technology to either reduce costs or revenue. One project I'm most proud of was for a bank I was working at. When I arrived, their infrastructure, software, and operations were archaic. They were performing so many manual actions not only in IT but throughout the entire organization. By the time we were done working together, IT had tripled in size, the overall employee count stayed the same, and the bank significantly grew its market share. Together, we were doing more than two times the work with the same amount of employees. They saved millions. If you follow the guidelines of this book, you can too.

Like most worthwhile things in life, you cannot succeed in this without work. You can't simply jump on a technology board and start surfing immediately. You will need some baseline understanding of what is happening before you even get into the water. For your situational awareness about technology, you will need to understand why right now is such a critical time in technology history. The next couple of sections will dive into why we are at a precipice of extreme technological expansion.

THE SPEED OF INNOVATION IS INCREASING

Technology has been a wave, far out at sea, slowly building momentum and power throughout most of human existence.

"Precisely how slow has this wave been building?" you may ask. So slow, it might not even look like it's moving. Don't believe me? Let's look back at early humans and the use of controlled fire and tools.

The ability to make fire on-demand is, in fact, technology. It might not seem like it to us, since it's something we take for granted today, but it is, nonetheless.

There is some debate in the scientific community about how long it's taken humans to gain power over fire. Some put it as far back as 1.7 million years, but almost everyone agrees that it was at least 400,000 years ago. However, the *match* wasn't invented until around 500 AD in China. So, it took at least 400,000 years for humans to master fire with a match. Talk about innovation at a snail's pace.

The matches the Chinese used weren't like the ones we use today, either; they were much cruder. Today's typical match, called a friction match, wasn't invented until 1826—nearly 1,300 hundred years after the Chinese match.

In its early days, technology moved like a glacier.

Another excellent example of this sloth's pace of innovation is the ax. The earliest axes, called hand axes, were sharpened stones used to cut and slice. Early humans used them as far back as 2 million years ago. The more modern ax (with a handle, used to increase leverage) wasn't invented until around 50,000 years ago in Australia. It took a monumental amount of time to advance something we see as simplistic. Incredibly, it took nearly two million years for someone to add a *handle* to a *stone*. Meanwhile, you and I learned about simple tools like this (our teachers called them "simple machines") in the second grade.

In the twenty-first century, adding a lever to a rock to make it more effective is, quite literally, child's play. Why, when it took millennia to develop, do humans now view that ax as simple?

TECHNOLOGY BEGETS TECHNOLOGY

Today, innovation happens at an accelerated pace. Technology begets technology, and everything we invent allows us to advance and develop more. We would have a very long conversation if we discussed how today's technology created more room for new technology, because modern humanity is so deep into this principle. So, let's start with an older technology, to show how it unlocked a new era of exponential innovation. Once we cover something simpler from a bygone era, we can dive into a more modern example. Let's power up the flux capacitor and take a trip to the Middle Ages. (You'll see what I mean when I'm talking about how the technology tsunami is building.)

THE BLACKSMITH

Many of us have seen a blacksmith working. Maybe you've seen one at a renaissance fair, or you got lost on YouTube one day, or you've seen a blacksmith in a movie. You may have a decent understanding of the tools of the trade—the hammers, tongs, coal, and anvil—and, of course, the cliché mental picture of some big burly dude with bulging muscles and sweat pouring over his body comes as a standard prop.

You probably think there isn't much technology involved with blacksmithing. After all, most of the tools being used are pretty simplistic. I doubt you've recently picked up a hammer at the local big box home improvement store and thought about what a technological wonder it is.

But it *is* a technological wonder. Like the ax, people were using stones as hammers for millions of years. But it wasn't until more recent history (around 50–30,000 years) that someone had enough brains to put a handle on it and increase its power. But let's go a little deeper with the technology behind blacksmithing— we aren't only talking about the hammer. We need to understand all the technology that goes into being a blacksmith. He needs a shop, a furnace, bellows for air, a chimney, a roof, walls, lighting, and a door, just to name a few. He also needs to power his *body's* internal furnace to keep his muscles doing heavy physical labor. Not to mention, he'll need a hefty water supply to stay hydrated in hot working conditions.

See how much technology goes into what we would, today, consider as a straightforward career? But we aren't done quite yet.

Remember, technology *begets* more technology. So, our blacksmith's resulting technology, say a sword, is the result of simple tools that took many millions of years to develop. Then, consider the blacksmith's shop... Can you imagine it? A creaky gray wooden building with a thatch roof, a door cobbled together, glassless windows with shutters. Inside, he has tables, a furnace, an anvil, and a chimney going up through the middle of the room. The whole room is filthy, covered with soot and dust from the dirt floor. It might look like something in *The Lord of the Rings* movies.

If you were to compare this humble setting against a house of today, you would consider it a third-world country. But hundreds of years ago, it, too, was a technological marvel.

This crude building was also the offspring of other technology. For it to be built, a significant amount of innovation and technology occurred. For instance, the ax and saw had to be invented before making the walls. Also, the logs had to be pulled out of the forest. To remove those logs, a horse harness needed to be crafted. To craft a harness, you need to know how to tan leather. You must know how to domesticate cattle to get their hides. That's quite a bit of technology to simply move a log.

A whole other technology set had to be created for the thatched roof. Seed had to be harvested, the land tilled with tooling, tooling that might be horse-drawn, requiring a harness. The seeds had to be sown and the resulting growth harvested with a scythe. Of course, the actual advent of the thatched roof had to take place as well. And that's just the building. Now, let's dive into the technology required to perform blacksmithing itself.

To shape metal into a sword, you need to heat metal to the point that it becomes soft and can be formed. For that, you need a forge that burns coal. But where did the forge come from? It had to be built out of stone and cement. Sure, no one invented rock, but they did create a way to shape it. It's held together with cement, another fantastic invention that we take for granted every day. (Of course, it's not the portland cement that we use today but an earlier recipe.) Finally, coal had to be discovered, and all the tools and technology for mining it needed to be invented before the blacksmith could even start his fire.

It's probably now clear why I chose a blacksmith rather than a more modern technology. Most of the technology discussed has developed recently, only being around for a few tens of thousands of years. Even with something as simple as a blacksmith, there had to be dozens, if not hundreds, of technological advancements taking place before the smith could make a sword. Millions of years of technology had to be developed for there to be a blacksmith profession. All along the way, each technological advancement laid the groundwork for the next. The blacksmith was able to create a whole host of capabilities that advanced mankind—from the metal tread of wagon wheels to advanced tools. The work of the blacksmith led to autonomous machines that could produce forgings at scale.

Because technology begets more technology, that means, at each new level, you're exponentially increasing the possibilities, and speeding up the development of the next innovation.

If we return to our tsunami analogy, the human discoveries of on-demand fire and crude weapons and tools were the first signs of this tsunami. These (and a few other early innovations) were the earthquake, if you will, that originated deep below the ocean and set off our wave. That wave moved very slow, gaining momentum for millions of years. By the time the blacksmith profession came along, the wave had gained significant speed. Humanity went from crude axes and hammers to swords in 30,000–40,000 years. It was quite an improvement from the millions of years it took to get to those first advancements in tools. But now we're just getting started.

The amount of technological advancement needed for a blacksmith is pretty impressive, especially since, by today's standards, it's considered crude technology. (I don't mean the output's crude, but the methods.) There are blacksmiths today that create refined pieces that are works of art, which require the workmanship of a skilled artisan, and a significant amount of time. With today's population, there is no way that the methods of a traditional blacksmith could scale to the level needed to supply the planet with enough knives. So, within this gap, technology advances. The need forces innovation to create at scale. Today, traditional blacksmithing is a fringe art, not a typical career for the masses. In the future, technology will push more and more jobs into the fringe.

It's not just individual jobs that are at risk. Entire industries will be changed. Take, for instance, the meat industry. There is a large upset right now about the amount of resources required to raise animals for meat. Countless companies are working out how to produce meat or meat equivalents at scale. It's very possible that soon, raising cattle will be a fringe business just like being a blacksmith. Unfortunately, your company could experience similar obsoletion, if you don't stay technologically relevant. Worse yet, a competitor can push you out of business.

THE WATER BOTTLE

One last example: the water bottle. No, not the fancy stainless steel ones, nor the ones with nozzles and spouts. I'm talking about the plain old "buy it for a dollar" disposable plastic water bottle. For us today, it's about as straightforward and unexciting as it gets. It's an everyday item that we dispose of without giving a second thought as to how much of a technological marvel it is. But let's dive in and see how much technology is required to make that water bottle.

To make a water bottle, first, we need a factory. A factory is made of steel, concrete, and glass. It would be pretty hard to dive into all these items to determine the technology required to make them, so let's just look at *steel*.

The steel in our factory is in beams, stairs, platforms, and corrugated steel for the roof. To have these items, you must mine ore, get it to a foundry via truck, melt it, form it, rack it, and stack it before taking it to a forge to be shaped into its final form.

Each of these steps may seem simple enough. However, each has its layers of complexity. How about the tires on the trucks and forklifts? Or the oil for the hydraulics on the excavator that digs up the ore? Each of these items has as many technological advances cobbled together as the building itself. Hell, even the *key* used to turn the ignition of the delivery truck is in itself a collection of dozens or hundreds of technological advances!

When it comes to the water bottle itself, we must consider what it's made of and how it's made. Bottles are made from various forms of plastic, either plant-based or oil-based. The first synthetic plastic, the great-grandmother of modern plastics, was called Bakelite, and it was invented in 1907. Bakelite was pretty brittle and didn't have the strength of modern plastics. It wouldn't work well for a water bottle. Most bottles today are made of Polyethylene terephthalate (PET). PET was invented in 1970, only sixty-three years after Bakelite. A trend should appear when we talk about all these technology innovations. First, it took millions of years for a hammer handle, then hundreds of thousands of years to invent a match. Plastics evolved from their first invention to modern plastic within a century. Can you feel the tsunami starting to gain momentum?

We won't go any deeper into the technology for the water bottle. If

I had started with an even more modern example, we'd be going down many, many rabbit holes of innovation, and it might take an entire book to cover it all. Yet, the idea should be clear: everything invented today is based on millions of years of technological evolution. With every evolution, the next comes faster. The wave is getting larger and more powerful.

And, it's unstoppable.

UNSTOPPABLE FORCE

The tsunami wave started to gain power with the industrial revolution in the mid-eighteenth century, when advancements in machinery allowed us to create products at scale. Consider textiles, where humanity was suddenly able to churn out miles and miles of fabric easily, compared to the by-hand methods of the previous millennia. Could you imagine how much clothing would cost today if it was still made by hand? Your local Target would hold half a billion dollars of inventory, and your closet would be empty due to the cost! Right now we are in the fourth industrial revolution, also called Industry 4.0. This revolution includes a knowledge economy, and it's driven more by intellectual capability instead of physical output.

During the industrial revolution, technological advancement got a significant boost, adding to the power and size of the wave. At this point, technology became a true force to be reckoned with. Still, the wave wasn't at the point where it was terrifying. If you were to travel back in time, you would still consider the 1800s an antiquated time regarding technology. Millions of years of advancement had passed, and the exponential growth of technology was becoming visible. The tsunami was visible out in the ocean, but it was still a tiny swell of water.

The amount of innovation in my lifetime (I was born in the 1970s) is astounding. It wasn't until the last century—or, more accurately, since the 1950s—that things started to kick into high gear. We can attribute the massive explosion in technological advancement to the computer. Computers make the heavy lifting easy, and crunching numbers, running simulations, and testing are no longer burdensome for inventors. The advent of the microchip and modern-day computers has allowed us to process more data and develop more innovation than we could ever do with our brains alone.

I recently had an eye-opening experience when I visited the Kennedy Space Center in Cape Canaveral. If you haven't been there, I highly recommend it. Seeing the rockets that helped get us to the moon was a breathtaking experience. (There is nothing quite as awe-inspiring as standing in a massive building looking up at the mighty Saturn V rocket!) When I looked at one of the command modules, the piece that the astronauts sit in and splashes down into the ocean, I was amazed at how archaic it looked. We landed on the moon in 1969, and from what it looked like, we used a steam locomotive to get there. Gauges, dials, and switches were everywhere on the command module. It looked like someone from the prop department got overzealous with the design for the next space movie. It's amazing how quickly our view of the world gets skewed. Today, with our Teslas and smartphones, the ease of use and modern technological capabilities have spoiled us.

There I was, staring at a rocket that landed on the moon just fifty-two years earlier, thinking about how far we've come in such a small amount of time. When I was a kid, motor oil still came in a can! Talk about a lack of user-friendliness.

So, what does this all mean? In summary, the advancement of technology is generally unstoppable. Sure, there were a few exceptions: during the Dark Ages, humans in Europe went backward from the indoor plumbing and public baths of the Roman Empire to dumping shit in the streets—certainly a step in the wrong direction. But now, the speed and force of our technological tsunami won't allow for such stalls.

When Technology Gets Lost

Yes, technology has been lost in the past. Unfortunately, a massive loss like that causes the technology tsunami to slow a little. Consider the great pyramids of Egypt. To this day, we still don't definitively know how the ancient civilization built them. There are theories, but they all seem to have holes, and scholars can't definitively agree on how they were built. Another good example is Damascus steel. We can make things *like* Damascus, but the actual recipe and process have been lost to history. When these and other methods get lost, technology takes a small step backward.

Thankfully today, there are people all over the globe. Even though we are hyper-connected, we are still loosely coupled. Being connected in this way gives us the best of both worlds. We get to learn from each other yet not be slowed by socioeconomic forces. For instance, Russia's current invasion of Ukraine is putting significant strain on Ukraine's ability to advance technologically. But Ukraine has already shared what they know with much of the world. Even if Russia were to somehow annihilate every

Ukrainian, their knowledge wouldn't be lost. You can say the same about nearly every other country. We won't have a Damascus steel situation again. I view this as a very positive thing. As a whole, the technological evolution of humanity is safe, even if one country regresses (or faces near extinction).

Overall, modern connectedness means that the wave of technology is gaining speed and size, and it's not stopping. It might get slowed down here and there due to geopolitical shifts, but in general, it's on a strong positive trajectory.

WHAT THE TSUNAMI MEANS FOR YOU

So, how does the tsunami, the speed of technology, and the exponential explosion of advancement affect you? Throughout history, people and companies have had enough warning to properly adjust to the changes brought by technology. Farmers got new jobs as the proliferation of tractors and other equipment arrived. Wagon- and carriage-makers could find new jobs as the car became prolific. Even the workers at Blockbuster found new jobs when they were put out of business by Netflix.

But advancements and mass displacements are arriving faster and with more frequency. How long can we continue on this path until technological advancement can't be kept up with, and people end up out on the street? The answer is not very long.

Let me say this: not all technology is going to kill jobs. Some advancements, like on-demand grocery delivery (think of Instacart) do a great job of creating new jobs. Before these companies, you would go to the grocery store to get your own groceries. Now, you can sit back with your smartphone and click away, and someone else will do that for you. You never need to worry about going to the store to satisfy your five-boxes-of-Oreos craving ever again.

But unfortunately, not all technology is like on-demand delivery, which benefits people's jobs, the grocery stores' bottom lines, and our general convenience.

Most technological advancements have the potential to disrupt current companies and industries. Today, it's not unthinkable that you could be a mobile app company running systems on-premises with a healthy and viable business only to witness a competitor with lower price points and faster release cycles using the cloud come onto the market and put you out of business in six months. If that timescale surprises you, then keep reading. And this isn't just about mobile application business—every firm is vulnerable to disruption if it doesn't pay attention to the warning signs.

That's why I wrote this book. I want you and your company to be successful long-term. Trust me—it's going to take some work to get there. Also, it's not the kind of work where you put it in once, reach some endpoint, then dust off your hands. Technological adoption is work that never ends. The tech tsunami is going to create a new world. Like COVID-19, it's a world you must learn to live in.

WRAPPING UP

We are living in an unprecedented era. The compounding effect of technological innovation is about to push us so hard we won't know what hit us. It's going to be like all of those race movies where the driver reaches down and presses the red nitrous oxide boost button: he gets thrown back into his seat, and it looks like his face is melting off as he rockets down the track. Technology begets technology, and we are about to press a similar button.

For those who are prepared, they'll be able to handle that acceleration. You will know what is coming and be prepared. Not only that, but you'll also understand that this wave isn't going to dissipate. It's going to get bigger as humans continue to advance technology. For a frame of reference, when I started my career in 1995, being in IT pretty much meant that you did everything. You were responsible for database administration, network administration, and websites (plus, others probably saddled you with phone system management as well). Today, each of these areas is so broad and deep with technology that they are all specialist fields, like the medical profession—when you go to the doctor, you go to a podiatrist for your foot and a pulmonologist for your lungs. Plus, each technological "specialty" is continuing to get broken down even further, with sub-specialists in each area—another sign that the tsunami is closer than we can imagine.

I know we covered a lot of ground in this chapter, but we needed to establish a baseline so you could understand how critical this moment in human history is. Up to this time, we've been able to wait to react, adapting our companies to the changes in technology after we missed the initial moment. However, we won't have that opportunity for much longer. Now, to future-proof our companies, we must learn to be proactive.

CHAPTER 2

The Tsunami Factors

I believe we are approaching a nexus, a point in time where so many advanced technologies converge that we will be genuinely blindsided by the speed at which things progress. Artificial intelligence (AI), data storage and processing, computing power, and rapid prototyping have all advanced in recent years. The combination of these advancements, coupled with the speed at which we can communicate and share data, provides a recipe for success...and danger. I call these specific items *tsunami factors*.

When determining if an advancement is a tsunami factor, I look for three major elements:

1. **Its usage must be prolific.** While most any advancement will accelerate other technology, if one innovation is cost-prohibitive or only a small subset of industries use it, I don't consider that one to have as much tsunami power as one that has general and widespread use.
2. **It must fundamentally allow for the acceleration of work.** Technology doesn't always lead to less work; sometimes, it

ends up creating more. To be a tsunami factor, an advancement must decrease or shorten effort.

3. **It must be commoditized.** When a technology reaches commoditization, it's already been used prolifically; so, there's enough feedback that no more significant iteration will occur. In a word, it's "stabilized," providing a solid foundation on which other technologies can build.

Over the course of time, there have been countless technologies that could be classified as tsunami factors. From processing steel to the concept that *pathogens* cause disease, not spirits or bad humors that must be bled from the body. Although it would be very interesting to cover all of these items, we don't have the time or space. Instead, I've focused on a few key items that are of specific importance to us today. Plus, each item we'll discuss meets all of the tsunami factor criteria.

There are also a number of newer technologies such as blockchain, gene editing, virtual reality (VR), and Augmented Reality (AR) that have the potential to be tsunami factors but haven't quite proven themselves yet. In fact, VR has been a hot topic as far back as 1994, serving as the backdrop of Michael Crichton's book *Disclosure* and the movie of the same name starring Michael Douglas and Demi Moore. However, VR has never gained the traction everyone anticipated. So, it doesn't meet the tsunami factor criteria. AR and blockchain and the other technologies mentioned in this paragraph are in a similar position, so we won't discuss those technologies either.

ARTIFICIAL INTELLIGENCE

It's hard to walk through an airport or look at a magazine without seeing an article or ad about Artificial intelligence (AI). It's all the rage. A computer is using AI when it is processing problems as a human would. Artificial intelligence is a blanket term that covers a number of different technologies and concepts. For instance, Natural Language Processing (NLP), the technology Alexa and Siri use to understand what you're saying, are forms of AI. The same holds true for Machine Learning (ML), which is a type of AI that uses algorithms to allow machines to learn and improve on their own.

AI is different from a rules engine, the predecessor to AI. A rules engine follows a set of rules; if A happens, then B should occur. Rules engines are a complex set of If/then statements cobbled together that mimic intelligence. For instance, with a set of inputs, a rules engine can help you determine the species of an animal. If you input "feathers," "bill," "webbed feet," and "less than ten pounds," the rules engine would likely suggest that the animal in question is a duck. If the rest of the criteria were the same but the weight were over ten pounds, then a good rules engine would suggest that the animal is a goose.

A rules engine can be powerful, but it has limitations. Since it can't think outside of its programming, there will inevitably be questions that a rules engine can't answer. For example, if we want the rules engine above to know the difference between a goose and a swan, a programmer would have to add in a new criteria and question if the animal is over twenty pounds. Further, rules engines are very repetitive (in fact, they're easy to work around, if you know how they will react). For instance, a rules engine always gives the same answer. If you have ever played an older video game with an "AI" or "computer player," you might have seen this in action: the computer player always acts the same way, so you can anticipate it and beat it.

True AI changes the game, figuratively and literally, because it can think. It isn't pre-programmed with certain answers. It's programmed with an equation to *find* the right answer. (This is an oversimplification, but we don't need to dive into the inner workings of AI, just the premise.) This level of capability makes an AI system, such as a modern video game, more like a human. An AI-controlled player will react to what *you're* doing, not just follow a preprogrammed path. This significant difference makes modern AI much harder to beat, because you can't anticipate what it's going to do. As you change your actions, it will, too. In order to win, you must seek out flaws in its algorithm and tactics, just as you would with another human player.

The AI used today is called *narrow AI*. This categorization means it's only programmed to do a specific action, such as speaking or listening to natural language. We don't call it *general AI*, which would suggest that it actually mimics human intelligence. For instance, when you talk to Alexa or Siri, one AI listens and interprets what you say. Another AI responds to you, even though it probably seems like the same AI is listening and responding. General AI would be able to do both, mimicking human capabilities.

Narrow AI has become immensely powerful, and it's changing our world in unexpected ways. Most people don't know that they're training replacements to their jobs.

Every time you talk to Alexa, you make her a bit smarter. Whenever Alexa doesn't understand something you said, a team at Amazon reviews this, makes corrections, and creates new training material for the AI. In turn, Alexa will understand that same phrase the next time you use it.

Also: Did you know that most times a website asks you to prove you're human, you're actively training an AI? When you visit a website that needs to ensure that you're not a bot, you'll often find a ReCAPTCHA, which will ask you to select images containing a specific item, like a crosswalk, bus, or bicycle. Google, who owns ReCAPTCHA, already knows what some of the photos include, and that's how it protects a website. But Google also throws in new images, and when you click on those, you are providing more training data to Google to make its AI better.

There is a multitude of jobs that could be displaced by AI, particularly those jobs that heavily rely on speech. Fast-food chains are working on new AI technology right now, for associate order-taking roles and other positions. Call center jobs are also at risk. The next time you go through the drive-through, you might be talking to a computer instead of a person. I've seen this first-hand. At AWS I worked with a soda company that was using AI to automate all of their drink ordering for stores and restaurants, removing the need for hundreds of call center staff.

AI will directly replace jobs, but it also helps advance technology itself. The technology solves complex problems that would take massive amounts of time with conventional methods. Today, algorithms solve complex problems that would otherwise take decades. For instance, rather than taking years to perfect a material recipe, AI takes only days to find a number of probable formulas. The human mind is no longer a limitation in our progress.

Until recently, we didn't have the processing power to make AI genuinely effective. It's been on the fringe of science for quite some time, because the hardware required was very costly. With costs high, the barrier to entry has been pretty steep, making it difficult for startups to use it. Only large companies, such as IBM, had AI at their disposal. With the advent of cloud service providers—such as AWS, Azure, and GCP—which will be discussed later in this chapter, AI and related technologies are easily accessible. Now all companies, large or small, can use these tools to advance technology, meaning, again, the rate of advancement is accelerating.

AI meets all the requirements of the tsunami factor criteria: Its use is prolific; every time you ask Alexa a question, you're activating AI, and this is only one example. AI also reduces effort, and by significant amounts when trying to solve complex problems. Finally, AI is commoditized, primarily by the cloud vendors; and specific AI, such as NLP, synthetic vision, and voice synthesis are all types of AI that are also commoditized.

DATA STORAGE AND PROCESSING

Have you ever wondered how much data there is in the world or how much is created every day? It's not very easy to pin down an exact amount, but at least 2.5 million terabytes (TB) are created per day. That's a huge number, so let's try and wrap our heads around it.

The Library of Congress is the largest library in the world. Its collection of books is estimated to be the equivalent of 15TB of data. So, in terms of data, humans generate the equivalent of 166,666 of these libraries every day. That's an insane amount of data. But data in its raw form isn't going to advance technology. It must first be analyzed and processed.

Today's equipment puts out vast amounts of data about how it's operating. Companies process enormous amounts of data from windmills to jet engines to refine designs and operating parameters. This analysis creates many tiny incremental innovations. When people say innovation, everyone's mind goes to significant advancements and jumps in technology. In reality, that's not how it plays out. The big jumps are made of hundreds or thousands of small, incremental changes. The analysis of data plays a huge part in how those changes take place.

Nearly every product you purchase today uses data in its design. It doesn't matter if it's the brake pads in your car or your fancy smartwatch. Why do manufacturers use data for their designs? Well, because we're well past the phase where trial-and-error is accepted in the first iteration of consumer design. You use trial-and-error when you don't know the result. Today, we either know how things will react or use data to predict how things will react. We can test theories using computers rather than carrying out experiments, hoping for the best. Once we have a good idea of what will work, we move forward with the approaches that have the highest probability of success.

A good example is how the designs for airplanes and cars are tested for aerodynamic efficiency. "Back in the day," designers would put a model into a wind tunnel and blow air with smoke past the design to see how the shape disrupted airflow; they'd make tweaks to the model, and continue with the trial-and-error approach. Today, all of this is simulated with computers, because we have all the data captured on how materials, shapes, and air density react with each other. We don't need to do these tests manually anymore. All the data collected from simulations must go somewhere, and technological advancement comes in to save the day. Today, we can store crazy amounts of data.

My first job was working at a small computer shop. I remember when the 1-gigabyte hard drive came out. Breaking the gigabyte barrier was a huge deal; I remember thinking at the time about how long it would take me to fill up that amount of space. "It is the equivalent of 694 floppy disks!" I thought. If you would have told me that two decades later I'd be "streaming" ten times that amount to watch a movie over the internet, I would have thought you were crazy. Then, not only would it have cost a fortune to store that amount of data, but I would have had to imagine a physically monstrous drive, as I couldn't have comprehended that much data coming through over the internet.

But here we are.

Now, it's cheap to store dozens of gigabytes, doing so takes little physical space, and not only does data stream over the internet, but so do meetings, music, directions, intelligent devices, and more.

I recently had to buy a replacement hard drive for my storage device at home, and I put in a new 18-terabyte hard drive—a drive 18,000 times larger than the one I purchased in the mid-1990s. Let's put that into perspective. Imagine that you invested $10,000 in the stock market, and twenty-five years later, it went up 18,000 times. Your investment would then be worth $180,000,000.

More storage space means more data can be collected. More data collected represents more analysis of that data. More analysis leads to better—or at least more—predictions. More and better predictions mean better and faster outcomes. The availability of storage capacity is the snowball rolling down the side of the mountain, getting bigger and bigger, eventually allowing for increasingly rapid innovation. But for innovation, we need more than just storage; we need some way to *process* that data. So, that leads us to computing power advancement, which we'll cover next. First, let's review and check the tsunami factor boxes for data storage and processing:

Is the use of data storage prolific? Absolutely. Does it allow for the reduction of work effort? Just think about all the accounting data alone that data storage and processing has speed up over the last seventy years. Lastly, is it commoditized? Doing a quick Google search, you'll find that you can get a 20TB disk for just over $300. I'd say that qualifies.

COMPUTING POWER

We can't get enough of our computers. We use them every day at home, work, and now, thanks to our phones, even the bathroom. Our unquenchable thirst for computing power originally lessened our brain's workload, particularly when it comes to performing complex math. Now, we also have an increased desire for stimulation, so we demand more movies and video games, and this has precipitated an unbelievable advancement in computing power.

It's not easy to determine when the first computer was invented. There are many ways to classify a computer. Do you consider analog computers the "first"?[1] If you do, the Antikythera mechanism—thought to compute the position of the moon and the sun—would be the first computer. Most think our ancestors made this device somewhere between 200 and 50 BC.

---

1   Slide rules, watches, and the German Enigma machine used as a code generator in WWII are examples of analog computers.

Most consider the first modern computer to be the Intel 4004 CPU, which came out in 1971. The Intel 4004 had 2,250 transistors. Computers are digital machines, and they essentially understand two states, on and off. A transistor is a switch that controls the flow of electricity. The switch is either on or off, understood by a computer as 1 for on and 0 for off, respectively. (You've probably heard of "binary" before. Well, this is it!) The number of transistors that a CPU has is directly related to how powerful it is, how many problems it can solve at once, and how fast it can solve them.

Most people sitting on the toilet, browsing Facebook, or checking their Instagram have no idea how much power they have in their hands. The phone that you carry around in your pocket is thousands of times more powerful than all the computer power used by NASA to land on the moon. Think about that for a second: something that we take for granted, NASA would have given an arm to have their hands on back in the 1960s. To put it in numerical terms: while the first "modern computer" had a little over 2,000 transistors, the A15 chip used in the iPhone 14 has 15 *billion* transistors. Based on the performance available on the everyday smartphone, it doesn't take much imagination to speculate how much computer power is available for supercomputers operating in 2022. The current fastest supercomputer, the Hewlett Packard Enterprise Frontier, can perform 1,685.65 trillion floating point operations per second. A floating point is a number with a decimal point in it (which is harder for a computer to compute than an integer). To put this in human terms, to perform 1 trillion of these operations you would have to do one every second for 31,688.77 years. So, by extrapolation, it would take you over 53,395,577 years to do what frontier can do in one second.

But you don't even need a supercomputer to solve many of today's problems. The cloud can combine the power of many at-home PCs and do an amazing amount of heavy lifting for companies that drive innovation. For instance, companies have done years' worth of computations in the cloud, analyzing genomics and cellular signal transmission in mere weeks, cutting time to market and accelerating innovation even more.

Just go back one hundred years ago. In the early 1900s, most data analysis had to be done by hand. Slide rules, abacus, and other technologies made computations easier, but they were still slow and error-prone, as is anything done by a human. Computers allow us to solve problems faster and with more accuracy. Those same computers are then used to make new, faster, and more intelligent computers, further perpetuating processing speed and driving new developments at an ever-accelerating pace.

I remember hearing an announcement in 2003 that researchers had finally loaded a complete genetic DNA sequence into a computer. It took thirteen years and cost $1 billion to complete the task! Today, any one of dozens of available companies can take your cheek swab, sequence at least part of your DNA, and let you know what potential diseases you might be at risk for and where your family came from—all for about $200—and they'll be done in less than five weeks. Today, companies load hundreds of thousands of peoples' DNA into computers annually. In the short time of nineteen years, we went from a technological marvel of storing one DNA sequence to storing millions of DNA sequences, like it's as common as buying toilet paper. By 2022, estimates suggest that at least 1 million people will have had their entire genome sequenced, and 17 million will have had it partially sequenced through services like 23andMe and Ancestry. I wouldn't be surprised if, in the next few years, DNA sequencing becomes a common medical diagnostic process. These advances in the DNA space have led to new capabilities, such as Clustered Regularly Interspaced Short Palindromic Repeats (CRISPR) gene editing—a technology that will change the world and healthcare forever. I can't wait to see what happens next in this space. Disease may be a thing of the past.

But it's not all about how powerful computers are. It's also about how cheap and accessible they are. When computers first came around, the only people that could afford them were huge companies, the government, and significant research institutions. Today, you can get a small computer called Raspberry Pi for about $25.

Computers are increasing the speed and power of the tsunami with accessibility, price, and power. Companies will have a more challenging time compensating for a lack of action as its speed increases. Plus, humans aren't going to stop innovating, which just makes the wall of water taller and taller every day.

It's really hard to dispute whether computing power is a tsunami factor. Computers are used everywhere, from your laptop to your smartwatch to even your new smart oven. I'm not sure you can get more prolific than that. It's also indisputable, because computers greatly reduce the amount of human effort needed for a variety of tasks. Try to do your job without a computer for a day and let me know how much extra work it took! Since computers are everywhere, they're also super inexpensive and commoditized. As we discussed, you can get a Raspberry Pi computer for $25; plus, it will fit in the palm of your hand.

SPEED OF INFORMATION AND COLLECTIVE BRAINPOWER

The speed of information drives innovation. It, too, is a tsunami factor. Back in high school, if I wanted to get some information on jet engines, or even something as general as the Hoover Dam, I would need to go to the library. I would walk over to the card catalog at the library, look up my subject, find what titles talked about the Hoover Dam, then go find those titles on the shelves. Finally, I'd locate the information inside each book using the index. It was a pretty labor-intensive process. Honestly, looking back, it kind of feels like I was living in the stone age.

Today, to find information about the Hoover Dam today, I simply pull out my phone, and within about ten seconds, I'll have a whole plethora of information, more information than the library would have had in the first place. The phone is only one pathway to that information, too. In my house, with all our Echo smart devices, I could also just ask Alexa for the answer.

Wonder is something that I used to *do*. I'd wonder how tall the Washington Monument is. I'd sit there, think about it for a minute, and then move on. I don't wonder anymore.

How far is the Earth from the moon?

How many genes does it take to determine human eye color?

How deep is the deepest part of the ocean?

Not only will a smart device tell you how far the Earth is, or how many genes determine eye color, or how deep the ocean is, but that device will also offer related information. For instance, it would tell you that the deepest area is in the Mariana Trench in the Pacific and it's called the Challenger Deep. I didn't even ask my smart device for that information, but now I know.

We live in an age of questions. Now, it's not as important to know the answers as it used to be. Now, knowing what and how to ask the right question will get you the answer, so knowing what to ask is more important. The speed of information is an opposing force to knowing the answer. The faster you get the solution, the less you need to know it. I think that's a great thing. This development allows you to focus on things you must know and not worry about the rest. You can always ask for the answer if you need to.

With basically all of the world's knowledge at their fingertips, innovators can focus on the crucial questions—the ones that don't have answers. If you want to make a new power tool, you don't need to worry about finding out about the thermal properties of certain plastics. You can focus on the *design* and how to get it working. Those are the critical parts. You can focus on solving the actual problems that your customers have. You don't need to sweat the small stuff.

TAPPING INTO THE COLLECTIVE BRAIN

The speed of information allows us to tap into the collective brain-power of the world. We, as humans, are our own type of collective supercomputer. We can do things that computers can't, namely, *create*. Some will debate me on this. They will point out that AI has created paintings, ad copy, music, and a host of other items. But these things didn't pop out of nowhere. Someone put those ideas into the machines. *People* created the algorithms, loaded the AI with sample data, and trained them. A person had the spontaneous idea out of nothingness that created that AI.

I believe that humans are the only ones that can truly create. Computers can only simulate human creativity. Will this always be the case? Maybe, maybe not. Computer power will continue to increase, and someday, we will have quantum computers that will completely change the landscape of computing forever. But will a computer ever be able to invent blues music? I don't know how it could. A computer can't suffer. Blues came out of an entire people's difficulties, and it has heart and soul—two things a computer can't have. When I listen to blues, I feel that suffering. I'm connected on a spiritual level. Again, how could a computer ever understand that?

We need humans, we require that collective brainpower, and thanks to technology, we can tap into that brainpower all across the world at a moment's notice. For instance, let's say you want to create new software for a phone that tracks exercise and blood sugar levels for diabetics using a smart device. You might need to talk to some experts in the field, such as the designers of the intelligent glucose monitoring device. In a matter of days, I could be on a Zoom call with those people. Technology has helped us tap into the power of humanity's collective brain, making the tsunami even greater.

We saw the speed of information and collaborative brain power in action with the COVID-19 pandemic. Researchers from around the world were collaborating to find a solution to the disease that was affecting so many people, so rapidly. If it weren't for this tsunami factor, millions more people would have died. The rapid development of vaccines was mind-blowing. To develop these vaccines just fifty years before the COVID-19 pandemic, you would first go to the library to do research, then, maybe, reach out to some contacts. You might even need to handwrite a letter to speak with those contacts. One hundred years before the pandemic, and you might even need to hop on a train to speak with someone. There may be letters and horses involved, as well as weeks and months of time. If you go back 150 years...well, let's be honest, I would have probably given up.

You can see why technology was slow to move forward, and why the speed of information has had such a significant impact on the rate of innovation. But not all technology and innovation are spurred on by computers. In the next section, we will look more at the *physical world*.

It's a bit different to measure information speed and brain power against the tsunami factor criteria. Even though it might not seem like it, everyone does actually have a brain. Brains are both prolific and commoditized. When it comes to the speed of information, it's estimated that 83.37 percent of the world's population has a smartphone in 2022. Since phones are the primary mode of communication, I'd say that speed of information meets and exceeds the tsunami factor bar.

3D PRINTERS, LASER CUTTERS, WATER CUTTERS, AND CNC MACHINES—OH MY!

"Rapid prototyping" is a term many people aren't familiar with. It refers to the ability to take an idea and build the first physical working version of that idea quickly. Having an idea versus a real object you can hold in your hand are two different things. The physical world doesn't always translate from a drawing quite as expected. With a working prototype, you can test your new idea, and iterate.

Rapid prototyping is relatively new. It's made possible by four significant technological advancements: 3D printers, laser cutters, water cutters, and Computer Numerical Control (CNC) machines. These devices allow you to take your idea, upload it to a computer, and quickly and cheaply create parts to build it. Although these devices serve different purposes, all have the same result when it comes to rapidly produced components.

Of all of these technologies, 3D printing is the one you've probably seen in action, as you can find 3D printers in many schools and even homes. They range from inexpensive to costing a small fortune, depending on the quality of output and the type of material. But the technology is advancing, and the materials used and the designs are both getting more complex. For instance, companies are producing rocket engines and full-sized homes using 3D printing technology! More recently, Bugatti has been 3D printing brake calipers out of titanium, and another company is making metal sex toys. Nothing is off-limits when it comes to 3D printing technology. I can't wait to see what else happens with this technology, as it will only become more influential.

If your company wants to make a new product, 3D printers are a gift. Let's say that your company makes displays for conferences. Your client wants a booth built. It needs to be light and portable, and because of their branding, it must have all sorts of weird angles and shapes. You need to make it out of aluminum to be strong enough, and of course you want to ensure everything fits together correctly so the client will be happy. You may draw everything up on the computer, but you still need to make sure that the pieces fit into the pipe quickly and easily, and everything looks right when assembled.

Before 3D printing, ensuring the piping fit together would have been a daunting task. But with a 3D printer, you could crank out some inexpensive plastic versions of your designs to test the fit of the most complex connections (or the entire build). If something's off, you could modify your design, print out a new version of the part or the whole product, and test again. Once everything's right, you could move forward with production.

You can use laser or water cutters to cut high-precision parts using lasers or intense water pressure. Similar to 3D printing, you can use these parts for prototyping or production components. Although some have limited three-dimensional capabilities, these machines will produce a much more refined and precision output than most 3D-printed components. Like 3D printers, laser cutters and water cutters can take designs from computers to production in brief periods. Some of these are so inexpensive, many people have them in their homes.

Finally, we come to CNC machines. CNC machines have been around for a long time. They're not only for prototyping but for full-on production components. CNC machines cut parts out of solid blocks of material. They are high-precision machines capable of producing high-quality parts and molds. Significant innovations in this space are commoditizing the machines and making them much more accessible. At this point, many CNC machines are just as cost-effective as home 3D printers. For instance, MakerBot, a popular classroom and home 3D printer, charges $1,299 for their most cost-effective model, the Sketch, at the time of writing. Similarly, a BobsCNC quantum mini CNC router will cost you $1,145 before options.

THE IMPACT OF CLOUD

The cloud has significantly impacted the acceleration of technology innovation, not because cloud technology has drastically advanced over on-premises technology, but because the cloud makes technology more accessible and dynamic.

There are three primary reasons innovation is faster through the cloud. The first is the simple fact that you can use pay-as-you-go. The second is the unlimited capacity of computing and storage resources. The third reason the cloud is so essential to technology is that you don't need to upgrade your equipment every time a new model comes out.

Let's deep dive into each of these reasons.

1. PAY-AS-YOU-GO

Before the cloud, any form of experimentation that might lead to a technological advance required investment in hardware. In my experience, most companies operate their infrastructure lean. (Well, lean enough to cover three to five years of use.) It doesn't take a PhD to determine that this inherently limits innovation. I can personally tell you from experience how this goes: You come up with a fantastic idea and proceed to tell management about it, and they get excited. They buy into the concept, but then the big question comes out: "What will it take to make this happen?" I had this happen to me. I had the concept of the cloud for financial services before the cloud was really a thing. I worked for a bank that served other banks, so we were in a prime position to offer our services through "the cloud" (though I didn't call it that). But there was a barrier to entry of around $350,000.

Right about the time you need that kind of money is when your great idea begins to die on the vine (this is where mine did). The second that you talk about financial risk and a bunch of equipment potentially lying around going unutilized, bosses and other stakeholders lose interest, fast. Now, this isn't a hard and fast rule. Every company is different, but how many companies out there have big, fat R&D budgets?[2]

Plus, the market value of computer equipment drops in price just as badly, if not worse, than buying a new car. The second you open the box, that equipment starts depreciating at an alarming rate. According to standard accounting principles, computers completely depreciate over five years. But in many cases, three years might be a stretch for actual functional use.

The savior? The pay-as-you-go model. With pay-as-you-go in the cloud, you only pay for resources as long as you use them. It's very similar to your electric bill. When you're running your AC in the summer, your bill goes up; when you don't run it in the winter, your bill goes down. It's not like you had to pay for a generator, running it no matter what. With pay-as-you-go, you're talking about an *operating* expense versus a *capital* expense. Whenever you buy an expensive piece of hardware, accounting needs to create a depreciation account and track that asset until it's disposed of—a capital expense. When accounting works with pay-as-you-go, a much smaller monthly or quarterly bill for the cloud is recorded as an operating expense every time a bill comes; and, there's no long-term commitment, thus no long-term accounting; when your company no longer needs the cloud, it can stop paying for it. So, with the pay-as-you-go cloud, there's no large

2   Research and development budgets.

cash outlay, making the business (and your bosses) much happier. So, your new, great idea? Now, it may just work.

Overall, pay-as-you-go benefits innovation by significantly reducing the apprehension for experimentation. You are more likely to get the reduced funding required to prove your business concept. This reduced fear helps to drive innovation faster. Having worked at Amazon Web Services (AWS), I can tell you that I've seen this first-hand, and the impact that it has had on innovation is astounding. If you think of it at scale, it truly is mind-boggling. Thousands of companies are running experiments that would probably never have taken place without the cloud, further adding to the power of the technology tsunami.

2. UNLIMITED CAPACITY

When I talk about the unlimited capacity of the cloud and how it accelerates innovation, most people look at me with an inquisitive look. It isn't inherently obvious how one has to do with the other. Most people frame capacity around doing an amount of work for an indefinite amount of time. Capacity in their mind is fixed. *I run my CRM software on 4 servers of hardware forever*, they think. I don't fault them; this is how most companies run IT and, therefore, their default operating thought process. It's time to break away from this mindset and think about how the unlimited capacity of the cloud can help you get work done faster.

For instance, let's say you work at a firm that wants to design better cell tower antennas for 5G. So, you determine that you need to process some complex equations on how 5G signals are absorbed and reflected by various materials. To do this on-premises, you would start by considering how much equipment you could afford and house. You would then consider how long it would take for those on-premises systems to process the equations. This timeline would lock in your rate of innovation. You can only afford so much processing power, you don't want more unused equipment lying around when you're done with the project, and you don't want to purchase a bunch of data center space to house everything. In the end, your time to market with your new antennas is at the mercy of the cash in your bank account and the space in your computer racks.

With pay-as-you-go cloud, the equation gets flipped around. Instead of spending $100,000 on equipment that will allow you to do the calculations over a year period, you could pay $100,000 for cloud resources that allow you to process the calculations in a month, as that $100,00 could be used for more and faster computing power.

With clouds, the question you start asking yourself is, *How long do I want this to take?* That type of thought process can drastically increase your rate of innovation and time to market.

In the end, the net spend on your calculations may be the same, maybe even a bit cheaper with cloud. The real benefit is that once these calculations are complete, you can solve new problems and focus on creating new differentiators in the market.

3. UPGRADING EQUIPMENT

The final reason that the cloud provides a faster innovation speed is that when you operate in the cloud, you can update your server equipment for more performance and cost-effective models every time your cloud vendor does. With the cloud, you can just shut down your server, upgrade the equipment, then turn it back on. Updates take seconds compared to on-premises, where they would take weeks. Without the cloud, you need to get quotes, wait for shipments, rack equipment, migrate servers... It's a cornucopia of time wastage. Plus, during and after the COVID-19 pandemic, equipment wasn't available, no matter what you were willing to pay for it. The upgradeability with cloud allows you to always use the very best equipment to get the job done.

Of course, it's not always about getting the job done faster; sometimes it's about getting the work done *cheaper*. Sometimes, newer equipment is slightly cheaper in the cloud, not just faster. The cloud vendors do this to entice people to switch so they can remove old hardware. When you can get current workloads for less, it allows you to refocus the difference in funding to new innovations. This savings lends itself to faster innovation, further perpetuating the technology tsunami.

THE IMPACT OF CLOUD COMPETITION

An interesting byproduct of the cloud is cloud competition: to compete for your business, the major vendors (AWS, Azure, and Google Cloud Platform [GCP]) have also increased their rate of innovation. Instead of a space race, we see a cloud race between the top vendors trying to capture more and more market share. This in turn results in more pay-as-you-go technologies being available to companies around the globe through their innovation and competition. The more technologies available, the greater the experimentation and rate of innovation.

Cloud competition is a major accelerator to the technology tsunami.

* * *

Measuring cloud against the tsunami factors is fairly straight-forward. Cloud is used by almost all companies in some form or fashion whether they are aware of it or not, so it's certainly prolific. Additionally, as we discussed, cloud advances technology and reduces effort in a number of ways. The pay-as-you-go nature of cloud, the ease of consumption, and cloud competition check the box for commoditization.

EXPONENTIAL DECAY OF TIME TO MARKET AND MASS ADOPTION

All the things we discussed—computing power, data processing, AI, cloud—have a marked effect on the time it takes to conceive an idea and get it to the market so people can start using it. In business, this is what is referred to as time to market (TTM). TTM is essential because the sooner a product is out in the field, the sooner you can identify the problems or necessary improvements. The sooner these issues pop up, the faster the rate of innovation. TTM isn't a tsunami factor directly, but rather a byproduct of them. However, it's vital to understand the role it plays in the tech tsunami.

We can go back to our ax example to solidify this concept. The first ax, the hand ax, was a piece of sharpened stone. Assume you're one of mankind's ancestors, and you're really good at making axes.

Later, you need some clothes, which means you need to hunt animals for their skins. However, you're not very good at hunting (when you need food, you gather). So, instead of hunting animals for their skins, you use a new rock to make another hand ax, and you *trade that ax with someone for one of their animal skins* to keep you warm next winter.

Trading is a good start to a free market, but here's why innovation takes so long in this type of scenario:

> To improve your ax, you need feedback. So, naturally, you'd like to ask your "market" how it worked. But a bear ate the poor sap you traded the ax to, so you don't have any feedback from them on improving your ax. Without feedback, you make your next ax the same as before, with the same flaws. Your ax might be made out of poor material, or cut the user's hand, but you'll never know, because you don't get feedback. So, even if you trade dozens of axes, they'll never improve much.

Mass adoption is key to innovation. You need feedback so that you can rapidly innovate. Remember, it took millions of years just for the ax to get a handle. Without mass adoption, there was no feedback loop, so it took a long time to get a better product to market. The exponential decay of TTM in modern times, along with mass adoption, directly influences exponential technology growth. Over time, thanks to the tsunami effect and the advancements in tooling we've discussed, TTM has become exponentially less, which lends itself to faster mass adoption. With mass adoption occurring faster, your feedback loop becomes faster, driving more and more innovation.

WRAPPING UP

Computing, data storage, AI, 3D printing, and the rest of the technologies that we discussed here have been around for some time. You might be asking, *Why are these tsunami factors?* Remember, tsunami factors need to be prolific and commoditized. It's taken a bit for technology to advance enough to get there. The first computers and hard drives were crazy expensive and could only be afforded by the government. Now, they're everywhere. The same holds true for the rest of the technologies.

The proliferation of the cloud highlights the tech tsunami's speed. The cloud went from an experiment by Amazon to being everywhere in just ten years. It reached commoditization in record time, and shows the ferocity at which new technology will be coming in the future.

The amount of compounding technology that we see right now is truly awe-inspiring. Much of this advancement is brought on by the rapid prototyping technologies we just discussed. Now that we have covered why this is such a vital point in human history, we need to cover what will happen to your company in the future if you don't take action now. In the next chapter, we will cover what has happened to companies recently to show you the warning signs of what is to come. Understanding the warning signs is the first step in being able to recognize what is going on in your company and actually *do* something about it.

CHAPTER 3

Heartache vs. Heartburn

Now that we understand the technology tsunami, we can focus our attention on what it looks like in the real world. Specifically, we will look at a couple of companies that have failed over the last few years and dissect what happened to them. We will compare and contrast the *heartache* of failure versus the better path of *heartburn* of change.

I call it heartburn because change is challenging, and many employees don't deal with it very well. But heartburn is necessary to avoid heartache. The companies we will discuss failed before the tsunami even made landfall. I want to give you perspective on the potential devastation that could occur if more companies don't adjust to the rapid pace of innovation.

Before we get into the company analysis, let's discuss a foundation concept, which will help you understand why companies struggle to adapt.

THE DEER PATH IN THE WOODS

The "deer path in the woods" is what I like to call the journey you will need to embark on as a company if you want to ride the tech tsunami. You may remember as a child, or perhaps hiking as an adult, when you once came upon a well-trodden path through the forest. The path was likely entirely devoid of plant life and leaves, and all the twigs and branches were broken or eaten off by the deer. It's clear the deer effortlessly run back and forth through the forest on the "road" they've made. That's your company right now: You and your staff know precisely what you are going to do. Your processes are set, and your products have been refined. In addition, you're making good profits, and life seems pretty awesome. You and your staff are at a point in your company's life where your brain looks like the deer path in the woods. Performing the same tasks repeatedly creates pathways in your brain for efficiency, so, the next time, you're more likely to repeat the same action the same way. Over time, you have a mental map that is hard to break. At work, you can perform many of the functions of your job without putting any real deep thought into it.

Think back to when you just started your career. You had a lot to learn, and your brain was fully engaged, making new connections and building new skills. You probably came home after work and felt utterly exhausted even though you didn't do any physical work, because your brain was working so hard. But over time, paths in your brain became more and more ingrained, so it was easier to perform your role. Now, you only need to use your creative thinking skills to solve a couple of dynamic problems every day. This repetition sounds excellent on the surface. Who wants to toil away every day, struggling to get things done because they're learning from scratch?

Many of the components of your job get downloaded into a primary region of your brain called the basal ganglia. You can think of the basal ganglia whenever someone says "muscle memory." (Muscle memory is a personification; your muscles don't have memories or think.) You will see the basal ganglia in action best when you're driving. Say you're driving home from work. When you get home, you may not even realize how you got there. Your basal ganglia put you entirely on autopilot. You were daydreaming in the car, thinking about whatever, but somehow, you were able to steer, sort directions, and get where you're supposed to be. Many components of your and your employee's jobs are downloaded in this same way.

Another part of your brain is called the prefrontal cortex. That's what you were using when you first started your role. That part is the engaging, creative area that you use when solving a problem that's novel. Whenever you take a second to think, you're engaging your prefrontal cortex.

You engage that prefrontal cortex whenever you learn a new skill. You must write new synapse connections. Unfortunately, that process is uncomfortable (which is why you were exhausted when you first came home from your new job). So, going back to the deer path example, what I'm asking you to do with your company is go into that forest, see that path you're on, and then choose to find a new path, to purposely get uncomfortable, to reengage that prefrontal cortex, as if you were new. If you do choose a new path, you'll run into branches, step in holes, and maybe even slam into some raspberry thorns. It's going to be uncomfortable. The good news is, if you push forward, you'll get to another part of the forest that you've never seen before. When you get there, it's going to feel amazing. You'll likely be covered with scratches, but those are your battle wounds, the parts that make you unique. You shouldn't shy away from them, as they are learnings that can change your life forever.

The best news? You'll use that new path you just forged, a couple more times. Eventually, it too will become well-trodden. Soon, your brain will become accustomed to your new path, and start downloading the new patterns to make them automatic and, again, you'll feel comfortable. It's a never-ending battle of finding new paths.

Overriding you and your employees' natural comfort is critical to future success. It's hard at first, but practice by taking your team and company off the usual path, with minor detours. That way, you experience a small number of cuts and scrapes, but it's not as drastic as creating an entirely new course. Wherever I've worked, I've tried to create an environment of constant innovation rather than a stagnant one with large changes made infrequently. I attempt to incrementally improve and keep change moving. I like the analogy of a car manufacturer. Car manufacturers don't drastically change a model. They continually make small changes to improve the design. A 2022 Chevy 1500 isn't going to be drastically different from a 2023. People are used to these small changes. So when the manufacturers *do* initiate a major redesign, people are ready and more accepting. The manufacturers have created a culture of continual change.

As we walk through the next couple examples, focus on the deer path concept and how these companies got stuck and ultimately failed.

THE REAL-WORLD

We have discussed a lot when it comes to the timeline of technology. But now it's time to put the rubber on the road. I've been talking about the tsunami, but I haven't told you what it will look like when it makes landfall. I want to paint a picture of how the future might look for your company. Although these examples are from the past, they foreshadow what will happen in the future. The companies we will look at in this chapter are Kodak, Blockbuster, and Sears. Let's look at our first real-world example of rapid technological displacements.

KODAK

Kodak was one of the largest and most pioneering companies when it came to photography and film. Not only were they in consumer film, but they were also in other segments, such as movies, TV, and X-ray film. Unfortunately, Kodak is no longer with us. Kodak was put out of business by digital photography, both in the consumer and the business-to-business market.

The increased capabilities of digital cameras took a significant chunk of Kodak's business. Obviously, digital cameras don't need film. They allow you to take thousands and thousands of photos without ever loading a single roll. This feature, of course, is what makes digital cameras so enticing. You can take as many pictures as you want, keep the ones you like, and delete the rest. With old-school film, throwing out unwanted photos would cost you a small fortune. Not only was digital photography great for consumers, but professional photographers loved it too.

As soon as I started using a digital camera, the number of photos I took significantly rose. At home, I have a box of pictures that contains about 200 or 300 photos that I had taken for the entirety of my life, until I bought a digital camera when I was about twenty years old. Today, my wife and I take 2,000 yearly. (Let's not even talk about my teenage daughters, who probably take hundreds of photos a day!)

Another benefit of digital photography is that you get to see your photos immediately; you don't need to wait for them to develop. This review capability is even more advantageous to professional photographers and radiologists. Plus, digital photos improve not only the capture of images but the storage, quality, and speed as well. Of course, not all of these capabilities were available when the technology was introduced. However, you can't just look at what technology does today—you must consider an innovation's long-term potential. (Later, we'll discuss the long-term view in more detail.)

Unfortunately, Kodak didn't jump on the digital photography bandwagon. They clung to existing film technology, and when their tsunami came, they couldn't get out of the way. Rather than riding the wave, they were swept away. It was unfortunate for such a long-standing American company. What makes the story even more gut-wrenching is that Kodak had spent more money researching digital photography than all of the other companies combined. They owned the whole trove of intellectual property and patents that Apple, Samsung, and Google swooped in to acquire when Kodak went under. You might even be shocked to hear that Kodak invented OLED technology, the advanced technology used in super thin TVs and the latest smartphone screens.

How could a company have all these patents around digital photographs and then be overtaken by that same technology? That's easy. They weren't able to see the tsunami coming. They didn't know how to get out of their own way. Management was so fixated on *film* that they never wanted to move forward with digital photography. They ended up with the heartache of failure, as tens of thousands of people lost their jobs.

You might be asking, "Where did Kodak go wrong?" Let's take a look at how most companies work. Kodak was no different, and you might be able to see where your company could fall into the same trap.

Perhaps like your company, Kodak had a bunch of sales representatives. Kodak compensated these representatives based on selling film to wholesalers and large retailers. These representatives were making a great deal of money, selling the film that they had been selling for years, and they also had deep relationships with their buyers. Their jobs were easy, and their commission checks were fat and happy. They were trodding down their well-known path in the woods.

Then came digital photography, which allowed someone to take photos, forever, without continually purchasing more film. To avoid heartache, Kodak would have had to choose heartburn, making the conscious decision to cannibalize its own sales for less revenue today on digital photography. Kodak had two major detractors from choosing this path. First, their sales teams didn't want to lose their commissions on film, nor did they want to learn how to sell something new, particularly technology. Secondly, management didn't understand the full impact digital photography would have on their industry. They saw a wave out in the ocean, but didn't understand its full impact. Kodak was a film company, and they would stick to it. Kodak spent billions of dollars on digital photography research and development but then failed to move forward by taking any of it to market in an effective way.

Hindsight is 20/20, and we all think what Kodak did was radically stupid. But that's because we already know the answer to the problem.

If we look at it, it's easy to see why Kodak didn't forge a new path. Branching off into new technology requires a significant amount of effort for an organization. You need to change the way you sell, change the way you advertise, and how you develop. Frankly, there's a lot of chance involved. But ultimately, to survive, that's what we must do. Not only do you need the foresight to understand the wave, but you must have the guts to follow through. You've got to go into that forest and tear your legs up on those raspberry bushes and branches. You'll need to twist your ankle on rocks and holes in the ground, because that's what it takes to move your company forward by finding a new path. (Plus, that's what's going to keep you alive and competitive.)

Imagine where Kodak would be today if they had continued researching and developing digital photography and utilized that intellectual property. Think about all the new technology they could have developed and where digital photography might be now. We might be looking at cameras that have ten times the resolution they currently do, because of Kodak's continued investment. Unfortunately, we'll never know, because Kodak decided it wasn't worth pursuing.

The moral of the story: Don't be Kodak. When you see what's coming, have the guts to get behind your idea and stick with it.

Even if that means potentially going to your shareholders and saying, "Hey, we're changing direction. Yeah, it will hurt sales in the short term, but we've got something newer and better for the long-term health of our organization."

What Kodak Could Have Done

This juncture brings us to the discussion of heartburn. What *should* Kodak have done?

It's not an easy question. At first glance, the answer would be something like, "Sell more digital cameras." However, as we discussed, Kodak had massive momentum behind its film—the sales team for one, and established marketing for another. That's a lot to undo to go in a new direction.

Kodak had a path through the woods to the film business. A better path for Kodak probably would have been to move all of the digital business to a new subsidiary, where it wouldn't conflict with the older business. This would have allowed different compensation structures and new sales teams. It would also have allowed Kodak to cannibalize themselves, rather than be taken out by the pack of competitive hyenas.

Yes, this tactic would have caused a massive amount of temporary heartburn. However, it would have prevented the demise of Kodak. Yes, the film side would have eventually been gutted (which has happened anyway), but the digital side would have taken off. Just a thought.

BLOCKBUSTER

The next company that we're going to talk about is Blockbuster. Most people are familiar with their story—the US king of video rentals. There aren't many people born before the year 2000 who don't remember going to Blockbuster on Friday night to pick up a movie to take home and watch while eating their pizza and drinking cold beer (or root beer, in my case).

Looking back, it's kind of hard to believe that a company with such market penetration could disappear within such a short amount of time. Most know that Blockbuster was derailed by Netflix, a seemingly swift destruction. However, it did take around *thirteen* years for that to happen (Netflix started in 1997, and Blockbuster filed for bankruptcy in 2010). Today, we talk about it like it happened overnight, but the truth is, Netflix was around for some time before Blockbuster went under.

At first, Netflix focused primarily on foreign and indie films. That's where they proved their business concept and the technology, packing, shipping, and tracking. They mailed DVDs all over the United States. Blockbuster was aware, but they didn't see Netflix as a threat, because they were selling something (foreign and indie films) that didn't interfere with their core business. Their business was big blockbuster movies from the big studios.

Believe it or not, Netflix tried to sell to Blockbuster for $50 million in 2000, and Blockbuster passed on the opportunity. Stop and think, just for a second, about how different the world would be if the founders *had* sold Netflix. Blockbuster probably would have scrapped the concept of mail-order DVDs. They probably wouldn't have seen the value and chalked the $50 million up to protecting their market share. They definitely wouldn't have gone into the streaming realm, and we might not have ever been blessed with *Bird Box* or *Stranger Things*. They didn't purchase Netflix, not understanding the threat, so how could they have possibly taken Netflix to the next level?

Blockbuster failed to see that Netflix would eventually transition from indie films to mainstream films. The jump wasn't hard to make, and the second that Netflix jumped, they captured massive portions of Blockbuster's market share. (But I'm actually thankful that Blockbuster never bought Netflix!)

Blockbuster also failed to anticipate that people would be willing to *wait* for their DVDs. They thought that everyone wanted them immediately. They misjudged their primary customers' needs. Blockbuster management was so accustomed to people going on Friday night to pick up the latest movie to then go home and watch it.

Blockbuster also didn't understand the power of the internet and how that would transform the way people watch TV and movies. Netflix didn't offer streaming until 2007, ten years after they launched. Blockbuster failed to see additional use cases of video rentals where technology, in general, was going. They didn't understand the threat until it was too late. Blockbuster eventually did try to come out with their own DVD mailing program, but the ship had proverbially sunk by that time, and there wasn't anything they could do to survive. So yet again, another cultural icon of America was destroyed by their lack of technological adoption.

Ultimately, I'm happy that Blockbuster didn't purchase Netflix because I don't believe we would be where we are today. If they had completed the sale, Blockbuster could have scrapped the new technology. Likely, they would have never invested in streaming, and as a society, we would probably be farther behind because of it.

Interestingly, Blockbuster's well-trodden mistake was drastically different from Kodak's mistake. Kodak was investing in the technology that eventually caused its downfall—they were researching and developing it. They spent around $50 billion on digital photography. Blockbuster wasn't investing at all in what would become their disruption. They didn't see technology as a way to improve or supplement their business. This lack of insight or even exploration will be increasingly dangerous in the future.

No new company would shy away from being the next titan-killer who takes out an iconic American company. Overall, there has been a major cultural shift in the way Americans see disruptors: Now, as a culture, our country *values* taking down a traditional company. So, it's unlikely that founders would sell once they have moved past the initial startup phase into disruptor status. Today, Netflix would likely not even go to Blockbuster with an offer, because there's pride in being the next David who took down a Goliath. This attitude shift puts additional stress on existing industry titans to remain nimble. Once a startup becomes big enough to show potential, they're no longer an acquisition target.

What Blockbuster Should Have Done

What's the heartburn path that Blockbuster could have taken? They should have shifted to an online model that ran alongside their physical stores. They could have easily done this by asking key employees to look at the marketplace competition—not just their direct in-store video rental competitors (such as *Hollywood*

Video), but the adjacent competitors, such as Netflix, whose initial success should have been an indicator of things to come.

If Blockbuster had done some competitor analysis, they could have *annihilated* Netflix. Not only could they have offered movies via the mail, but they could have also allowed for reservation and local pickup, something Netflix could *not* do.

Furthermore, with some careful observation, Netflix would have seen the capabilities of Redbox as well. Think of how powerful Blockbuster's model would have been if they had offered all three: mail, kiosk, *and* in-store rentals? Think about it: We could have rented a movie from any location and returned it to any location. It would have been the ultimate in convenience and market saturation. Blockbuster wouldn't have even had to be as cannibalistic as Kodak to survive; likely, their stores would have slowly closed due to declining traffic, but some still would have stayed around for the long run. Heartburn, yes. Heartache, no.

Through my consulting career I have helped numerous companies avoid the heartache of failing to adopt technology and adapt to the tsunami. Banks, media companies, and even a number of small businesses are on the list. But unfortunately, one of the biggest setbacks that I face in trying to help these companies is their resistance to change.

One such company that comes to mind is a very large financial services firm that I was assisting. They had been recently hacked and wanted to move to the cloud to get a better security posture. Unfortunately, they had some business processes in place that were inhibiting the correct adoption of cloud technology. They faced a significant amount of heartburn working through improving those processes. People were very hesitant to change. In the end, everything worked out, and the modified processes worked effectively. If they hadn't changed, they would have run the risk of yet another breach. Another public relations hit for a second hack would have been devastating to their business.

SEARS

The last company that we're going to discuss is Sears. I bring up Sears because it's a unique story in that the company has essentially been put out of business by technology that mimics a process it pioneered.

You might be confused about why we're talking about Sears because, technically, Sears isn't out of business. But as far as I'm concerned, it's dead. They just don't know it yet. They're suffering, walking around like zombies in purgatory.

Failing to adopt technology wasn't Sears' only issue (there was also a significant amount of mismanagement), but it was their core problem. Originally, Sears lost some traction to other big box stores, such as Walmart, but the death blow came with the internet. Amazon has dealt significant damage to Sears, and this is where I want to focus.

Richard Warren Sears started the company in the 1800s. The firm essentially pioneered the catalog, delivering products straight to your home. Thanks to the internet, catalogs are virtually extinct today. However, they had a long and successful run, thanks in part to Sears.

Amazon touts itself as the everything store. You can buy pretty much anything on Amazon—a book or even a sauna. Ironically, Sears used to be just like Amazon. There wasn't much of anything that you couldn't buy from the Sears catalog. Of course they had clothing and houseware, but you could even purchase items like house kits (all of the supplies and plans needed to build a new home). Sears was the everything store back in the day. You still can't buy a house on Amazon, yet.

Essentially, Amazon is an updated version of Sears. Sears provided an easy way to source items through their catalog at a good price. It was a radically different approach than that of their competitors, and they swallowed up the market. Amazon mimicked this model using the internet. But for some reason, Sears was unable to make the next technological leap in order to survive.

A couple years back, I had a Sears shopping experience. I went to the store and proceeded to purchase a new dishwasher. I got to the checkout and found that the register was an antique. It was a plasticky yellow, and it looked like it had smoked fifteen packs a day for the last twenty years. After I paid, the associate handed me my receipt—with more information on it than an encyclopedia. Do you think that they could have emailed all that delivery and warranty information to me? Nope! I didn't even bother asking. Instead, they printed out what must have been a fifteen-foot-long dot matrix roll of four-inch receipt paper. Mind you, this was in 2016, not in 1995. Some people might think: *Sheesh, Aaron. It's just a receipt. What's the big deal?* But here's the issue: when companies don't keep their most basic technology updated, that points to a systemic problem. It demonstrates that companies are focused on bottom-line savings versus top-line revenue generation. Companies get into trouble when they start thinking like this. You can't *save* yourself billions of dollars in revenue. You can only *make* billions of dollars.

You see, everything boils down to customer experience, because customers buy things. Customer experience may not pay the bills, directly. However, when someone has a horrible experience, they'll never return. (Trust me: Next time, I'll be going to a store that automatically looks up my information and sends me an email confirmation and text messages when my delivery is coming.)

Sears is another example of when management wasn't looking at adjacent competition and how they were changing the overall market. Amazon started selling books, which probably didn't look like a threat to Sears. But that's how all disruptors start. It's like the movie *Jaws*. First, you see a fin and think it's a dolphin. Before you know it, a twenty-five-foot shark is chomping down your ass like you're in a game of Pac-Man. Then, it's too late.

We should always be on the lookout for adjacent threats and compensate.

What Sears Should Have Done

Things get a bit complicated regarding the heartburn that Sears should have endured. One of the issues is that Sears was contracting and working on cost-cutting strategies instead of trying to grow top-line revenue. If they had been focused on top-line growth, they would have naturally considered expanding deeper into the online market. But then, the infrastructure would have been costly to deploy and ramp up. This barrier to entry is why companies must have an R&D budget specifically earmarked for technology and innovation. Having a technology R&D budget allows a company to expand top-line revenue even when their market share is contracting due to competition. Of course, you can't steal the R&D budget away whenever you need it to pay for other things in the organization. With that strategy, you'll end up in a tailspin, unable to recover. Sears should have focused on assessing their business weaknesses and innovating to overcome them. Other retailers such as Best Buy and Target have made adjustments to their businesses to survive in a post-Amazon world. It was totally within Sear's power to do so as well.

WRAPPING UP

All three stories here show that these companies didn't compensate for the changes in their environments. Kodak, Blockbuster, and Sears all had good ideas. There were also people within those organizations who probably knew what the companies *should* have done. Most of these firms had the funding to pursue new technology. But, they got caught like deer in the woods, traveling too much on a well-worn path, unwilling to venture out.

In the rest of the book, I want to give you a path to prevent these kinds of futures in your own organization. There is absolutely no reason why your company needs to end up like the three of those. However, it's going to be even more critical in the future to act swiftly, because the tsunami isn't slowing down. Unlike Sears, you aren't going to have fifty years to fail. The old patterns are too slow, too cumbersome, and they're ultimately going to hold your company back rather than take it forward.

In the next chapter, we'll dive into the first step in saving your company from a similarly dismal future. We will look at the mentality you must adopt to become more nimble. Without your mind being in the right place, there's no hope for you creating the right collective mindset.

PART II

How to Surf

CHAPTER 4

Your New Mindset

So far, we've covered what the tech tsunami is and what happens when you ignore the changes in technology—potentially missing the impact it has on your organization. Now we're going to focus on how you can be ready so that you can prevent your company from suffering the same fate as the Kodaks of the world.

In particular, having the right mindset is critically important. The companies we talked about in the last chapter had a lot of lead time. Every one of them had at least ten years to adapt; they could have slowly developed a new mindset. Unfortunately, the tech tsunami will not afford you the time those companies had. Our new timescale is in months to years. Not decades or thirteen years like Blockbuster. Even a nimble, smaller company could be displaced by a larger one, like Apple: Several child screentime control applications were decimated when Apple pulled them from the app store and released its own native screentime application, forcing the other competitors out. In any case, you must adapt quickly. It's critical.

First, we must start thinking in a new way to remain competitive. Secondly, we must deal with significant mental blockers; we'll cover those later. For now, let's get started on your new mindset.

YOUR NEW WAY OF THINKING

Mindset is vital if you want your company to remain nimble enough to avoid obsolescence. You won't be able to help your team unless *you* have the right mindset. You are the leader, the person whom people will look up to and rely upon. Don't be part of the common trend in which managers, owners, executives, and board members don't understand technology.

Today, you can't rely on intelligent people under you to have all the answers. You must *know* where you need to go. Of course, you should always utilize the diverse capabilities of your team. However, ultimately, the ball is still in your court, and you must have a firm understanding of how technology will affect your business. You simply can't delegate something this critical. No, you don't need to be the one who knows how to implement all the technology. But, you should have enough information to make informed decisions *about* technology.

HANG 5: THE 5 TENETS OF TOMORROW'S SURFER

I don't want you to become what I call a "magazine manager." Magazines are a bit passe, and that's the point. A magazine manager sees the latest technology in a magazine and then barks orders, touting that the company "needs" it. Truthfully, you need to be informed, not sold on following the latest buzz. Sorry, but you're going to need to do a bit more legwork than a magazine manager.

We will cover the five key tenets executives must follow if they want to lead their organization into the future. These will help you get your mind in the right place. What's the right place? Let's look at our good friend Louis Pasteur.

Not only is Pasteur famous for inventing pasteurization, but he's also known for the quote "Chance favors only the prepared mind." Why does this matter? Well, because many people mistake *luck* for *preparedness*.

When someone makes an incredible stock play, what luck? Well, perhaps it was the hours and hours of research they put into understanding the market and its influences. When the opportunity presented itself, the investor was prepared. You must build similar mental preparedness.

I'm going to be upfront. Developing a new mindset is not going to be easy. You'll need to undo a lot of your brain's current hard wiring. Get uncomfortable, and get off the deer path. Like they say at the gym, "If you're not feeling the burn, you're not improving." Oh, and these tenets are not just for you. They are for you and your staff. But before you instill them into others, it's essential that you understand them and demonstrate them.

1. THINK LIKE A CHILD

The first tenet is what I call "thinking like a child." Return to your childhood, or when you had children of your own. Consider a child's thought processes, particularly when they're very young. There aren't many pathways engraved in a young child's synapses yet. They just really don't know that much. So, children don't have many biases. The few they do have are based on physical stimuli, such as, "I won't touch that hot curling iron again because it hurt last time." Those are good biases. Unfortunately, as we age, we introduce numerous other biases, and most of them harm our ability to think creatively. As you age, you become wound up in all of the "rules" of the world, which prevents you from expanding your viewpoint. And that's the mindset we want to get rid of. What rules are we talking about? Rules that have no basis in fact. Some examples:

- The earth is the center of the universe.
- The earth is flat (sorry flat-earthers, you're wrong).
- You can't travel faster than the speed of sound.
- No one can get to the moon.
- Humans can't fly.

These rules are some of the more grandiose, but I wanted to illuminate the absurdity of many biases. Faulty biases lean towards one outcome, not because of facts, but conjecture. When I'm working with organizations, I often run into faulty biases when working with technology adoption. They manifest in different ways. Here are some common biases that I see:

- "We can't do that because of [insert arbitrary reason not based on fact]."
- "It's too expensive."
- "We don't have the budget for that."

The bias I see the most is the "We can't do that because of X" bias. X is usually some lazy-ass, bullshit excuse. This bias arises because people are so used to their usual deer path that they can't even imagine getting off it. They are super comfortable and have no intention of stepping off the trail. So what do they do when faced with new ideas? They immediately throw up barriers. They start explaining all the reasons something can't be done.

I've been haunted by these biases in real life. At one company where I was working, they had a mix of different printers, some as old as twenty years. I saw an opportunity to revamp the entire fleet and save money and increase performance. I was met relentlessly with the bias that it would be too expensive; there was no way I could upgrade a ten-page-per-minute printer with a fifty-page-per-minute printer and save money. They even told me not to "waste my time" getting quotes. Thankfully, being the technology driver that I am, I didn't listen and got quotes anyway. Not only was I able to save money, but I was also able to get the printers under a management program so we never had to touch them or the messy toner again. What the others had failed to take into account was that the old printers were so antiquated that the toner cartridges were insanely expensive, several times the cost of newer cartridges.

I can assure you that you will find yourself automatically throwing up the "It's too expensive" bias while you're learning how to think like a child. Don't fixate on this mistake. Just notice it, and move on.

Unless your company is on the bleeding edge of technology, creating hyperloops, or colonizing Mars, you are not a snowflake. What do I mean by snowflake? That your company is so unique that not another one in the world has the same problems, conditions, or objectives. Unless you are a pioneer creating a state-of-the-art technology, the technology you're adopting has been used by someone else before you. If some company out in the world can use it and has worked around the blockers, so can you. Quite simply, if a vendor can sell their product to more than one company, then there is no real reason that you can't use it. You're going to need to let go of biases to surf the wave. There's no way around it.

And biases don't just manifest with mental obstacles; they can also manifest when you don't *like* a particular technology: "I don't like TikTok, and therefore, TikTok can't help our business." Unfortunately, today's clients age out, and new generations emerge. These generations may love using TikTok, and it's one of their primary social media tools. At the time of writing, nearly 50 percent of TikTok users (around 600 million people) are under thirty years old. If you're not going to move forward with TikTok, you're leaving millions of potential clients on the table. You can't afford to have bad biases. You must look at the world as a child, seeing the world for what it is. When a young person sees TikTok, they see a fantastic app, and so should you.

Another significant aspect of thinking like a child is asking "why?" a lot. Most people are afraid of asking why, perhaps because they're scared of looking foolish, or worse, they think they don't need to ask why, because they have a complete understanding. Either mentality can be devastating. You must keep asking why. Be willing to dig as deep as you can. You can't be afraid to look stupid.

As a leader, you don't need to be scared that you don't have all of the information. You will never have all the information needed to make a perfect decision. You must get comfortable with having *enough* information to make a decision.

On the flip side, you don't want to use bias to fill in the blanks either. Children are OK with admitting they don't have enough information to make a decision. But, as adults, we lose this comfortability. Instead, we use bias to fill in the blanks and make poor decisions. Again, combat this tendency by asking why. Be that four-year-old who won't let up. Let's put on our snorkel gear and dive a little deeper into this with a scenario.

Imagine that you're at a quarterly meeting. A key player on your team comes forward and says they want to adopt new software that helps email marketing. You ask the question, "Why should we buy now?" They should be able to answer this, if they've thought it through at all. If they can't answer the first why question you have, then maybe they shouldn't even be on your team.

Hopefully, they will have an answer to this first layer of questioning. Likely, it will be something around the lines of, "Right now, we send emails manually using our mail client and mail merge, and it's a clunky process." Here, most people stop asking why, which is a horrible idea. They quit, because the answer seems to make sense. You may biasedly assume that your team member has investigated the idea and thought it all the way. I'm not saying they haven't or that you shouldn't trust your team. However, it's your duty to ensure that they or a team member has considered all the outcomes and you or they have explored all the options. Here, don't let your biases take control, or theirs, for that matter. It is *your* duty to ensure the solution is the right one. So, ask why again:

"Why are we sending so many emails?"

Your team member may respond with something like, "For the last three years, we've been marketing through email, even though it takes a lot of time. We've had some success with it: a 10 percent click-through rate. And, we believe that if we're able to get out into the market with more emails, we could drive more sales."

Ok, that's a valid answer. But now, here's where your *new* mode of thinking kicks into high gear. Put a spin on the next question. Ask something along the lines of, "But, why are you using email?"

Even if they answered the other two satisfactorily, 90 percent of the time, they won't be prepared for *this* question. More than likely, you'll hear a response like, "That's what we've always done." And *that* is a bad answer. It means they haven't looked at other technologies. Every conversation should involve technology, even one seemingly as simple as this one, because it could lead you to a colossal technology play.

Now, it would be great if you got a unicorn answer in return, like, "We've tried LinkedIn, we've tried TikTok, and we've tried push messages through our mobile application, and we've found that email has the highest success rate." That would be a *stellar* answer. Unfortunately, you're probably not going to get close to that.

In every situation, keep digging deeper and root out the biases. Keep pushing until you get hung up. This is how you start exercising tenet one: think like a child. You ask why, every chance you get. If you do, I guarantee you'll start making better technology decisions today. In the above scenario, if you kept digging, you could find out that you want to change from email altogether, or you may discover that 10 percent isn't enough for your email campaigns anymore. Or maybe you discover that you not only want the suggested software, but that it can get you higher than that 10 percent success rate when used properly. I don't know, but you can if you ask "why?"

2. PROVE YOURSELF WRONG

The next tenet we're going to discuss is around proving yourself wrong. If you were to go back in time to about 1995 and ask me what the biggest hard drive available for computers was, I would have said that the biggest hard drive was one gigabyte. Back then, that was a fact. However, that "fact" did not stay a fact for very long. Hard drive capacity increased quickly. In 2022, twenty terabytes is the current maximum size. Today, facts are transient when related to technology. What's true when I write this won't necessarily be true by the time you're reading it.

One factor playing into the technology tsunami is that people get hung up on "facts." A statement may be a fact for such a fleeting amount of time that it's not even worth our brain power to remember it. I don't even bother storing information like maximum hard drive size anymore. As technology increases, facts are going to be more and more transient. So when I say prove yourself wrong, I mean precisely that. When you sit down, and you're having a conversation, and you feel yourself about to pull out one of those "facts," like "The largest instance in AWS has 4TB of RAM," I suggest that you question it first. Ask yourself, *Is this still accurate? Am I going to unintentionally lead my team down the wrong path because I'm using outdated information to make my decision?*

Be willing to constantly weed out inaccurate data. The speed at which your knowledge becomes obsolete is mind-boggling today. Start proving yourself wrong. Get in the habit of checking and rechecking information to ensure that you're making decisions with the latest and greatest data. Accurate facts could be the difference between the life and death of your company. Let's take a look at what lousy information might look like today:

Let's say, your company wants to develop a new application. You want this application to be cost-effective from a management and infrastructure perspective. If you're going to keep your infrastructure costs low, you "know" that it's a "fact" that a compiled computer language, such as C++ or C#, will be the fastest. The faster and more optimized the language, the less hardware it takes to run it, right? (A compiled language optimizes language by taking the code entered by a programmer and converting it into fast machine code.) So, C++ makes sense, because it will be the fastest, and will use less hardware, which will make it cheapest to run. To you, that's also a "fact." But before you use all those facts to make what seems like a straightforward decision, try to prove yourself wrong. And for good measure, sprinkle on a little bit of thinking like a child as well.

So, first, start with a "why" question:

"Why do we want to run a compiled language?"

Answer: "Because we want to save money on infrastructure costs."

A child wouldn't stop there. They'd go further:

"Is saving money on infrastructure the only place where you could optimize costs?"

Answer: "No, it's not. We could also save costs on engineering, maintenance, quality assurance, and in a whole host of other places."

Today's number-one expense for an application isn't in the processing power, but in the engineers' salaries. Once figuring that out, you'd have a new path to investigate:

"Is C++ really the most cost-effective language to write in?"

Answer: "When considering the engineer's salaries, no, it's not. Other languages, such as Python or Node.js, are easier to write code for than C++, which would decrease programming costs."

See? By applying tenets one and two, you found an answer other than what seemed like the most obvious one at first glance.

When it comes to writing programs today, the most expensive piece isn't the computing power, it's actually the personnel required. But thirty to forty years ago, when we talked about expensive mainframes and the original desktops, computer power *was* the most expensive. But now, computing power, especially with the cloud, is relatively cheap. So, the better course of action is to actually use a non-compiled language. You can trade the slight additional cost of infrastructure to save on the more expensive cost of engineering time. Plus, your time to market will reduce, because your engineers don't need to write as much code. So, by diving deeper, you're at a significant advantage because now, you've added the benefit of speed, allowing you to outrun your competitors.

Of course, this example is oversimplified on purpose, as there would be other factors to weigh. Still, it demonstrates that you can't take facts at face value, and you need to dig deeper and try to prove yourself wrong.

You might be wondering, *What if I can't prove myself wrong?* In that case, congratulations: you're right. The goal isn't truly to prove yourself mistaken, but to ensure you've exhausted all possible avenues. If after trying to prove yourself wrong, you're still right, you're on the best path.

3. ALWAYS BE CREATIVE (ABC)

The next piece to consider for your mental journey is ABC. Yes, I've seen the movie *Glengarry Glen Ross* as well, but I'm not talking about "always be closing." I want you to "always be creative." In our modern technology age, the most creative people will win. The ones who can solve problems in the most creative yet simplistic ways are the ones who are going to get the cheese, plain and simple.

First, it's important to create elegant and *simple* solutions. For instance, if you can create an algorithm that solves a coding problem simply while your competitor has built something complex, their solution will take longer to build, meaning it's going to take them longer to get to market. Simplicity equals elegance and speed, which leads me to the second point: As a society, we want things faster. We don't want to wait months. We want it now. Our whole society is based on now. After I pay for that car (in a frictionless manner, by the way), I want it in my driveway today.

If someone's waiting for you, a competitor will come along and steal your business. One of the best ways to defend against faster competition is to always be creative.

Lastly, differentiators will be the term of the future—both differentiating features and products. Differentiators are features or capabilities that give you a competitive edge—something that gives customers an additional value add that they want. Let's say you sell an app that tracks a company's budgetary spending, and you have many competitors in your space. You could create an AI algorithm that can anticipate future spending based on previous history and alert users of potential cash flow issues. With that feature, your app would no longer be just a reporting tool of what *was*, but a predictive mechanism of what *will be.* This gives you a differentiator in the market because you're the only ones (at the time) offering this feature.

Companies that can create those products and features and get them to market faster will be the clear winners. Unfortunately, most of the time, companies are looking at what the competition is doing; they aren't being creative. They're constantly playing a game of catch-up. So you and everyone else in your company will need to be looking for the next best thing constantly to identify those differentiators. Being content with where you're at isn't going to fly. So how are you going to stay on top of change?

You need to get creative, which is easier said than done. There isn't a silver bullet, but I'm going to give you a technique that has been very successful for me and others that I have mentored. First, we need to kick fear to the side. At one of the offices I worked at, a sign on the wall said, "Create something today, even if it sucks." I think this is an excellent mantra to keep thinking about. The fear of coming up with a dumb idea is very real. I've had many dumb ideas—so many, I can't even count. But that's OK, and it's part of the process. Make dumb shit; it helps to highlight the gems.

My Three-Step Process for Torching "The Box"

Now, onto my recommended process for torching the box. Everyone always talks about the proverbial "box" that we keep our minds in, telling people to "think outside the box." I'm saying that the box doesn't exist. It's a set of biases that we've constructed in our minds. Therefore, each of our minds is trapped by our own mind. The box is a zen riddle, a paradox. We will incorporate all of the other mindsets discussed so far to set this box on fire and get rid of it once and for all. To do this, we need to flush our biases and think like a child. We are certainly going to need to prove ourselves wrong. By incorporating all of these thought processes, we can truly get creative.

The first step is to truly understand the problem. We're talking about significant problems; you will not use this process because you ran out of coffee in the break room. Identifying the problem seems simple enough. But most of the time, people get it wrong. People are great at identifying symptoms but not root causes. You might have pain in your ankle. You recognize pain as the problem, so you take painkillers. But pain isn't the problem. It's a symptom. The real issue is that your ankle is broken. Fix that, and your pain will go away. For our scenario, the problem we're facing is that users are abandoning our mobile application. Yes—that's a symptom, not the problem. What we really need to determine is "why" users are leaving the application. We'll get to that.

The second step is to find all of the players involved. By players, I mean all stakeholders, such as your salespeople, managers, developers, customers—anyone impacted by the problem or impacting it. Depending on your organization and problem, there could be dozens of actors. Make a list if you need to. With the list of actors, you can move on to the next step.

The third step is to start looking at the problem through the eyes of all the players to uncover the root problem. Start with the player you think is closest to the issue; that's probably the users in our scenario. Then, once you uncover that player's root problem, take that issue, and go on to the next actor. For us, that would be those who are choosing to cancel their subscriptions. Look at the problem through each player's eyes, beginning with the player closest, putting yourself in their shoes. Once you've identified one actor's root problem, move on to the next.

So, let's do a little acting, and walk through all three steps with our scenario:

Step one: Identify the one large problem. Here, the problem is that users are not happy with the application and are therefore canceling the subscription. The underlying problem is not that they're canceling, or that revenue is decreasing, or any other surface-level symptom.

Step two: Identify the players. In this scenario, we have the users, developers, the development manager, the CTO, and the CEO.

Step three: Find the root issues for each actor. Starting with the actor closest to the problem, take their root problem to the next closest actor, drive to that root, and continue outward. (Full breakdown of step three below.)

For our scenario, we will say that the root problem from the *user's perspective* is that they want to use the application without internet access. The application is popular for taking photos while on vacation, and frequently people have poor access. So, needing to be online is causing users frustration, and they stop using the app as it doesn't fit a significant need.

From there, we move on to the next player, the developers. Now, you must imagine that you are a developer working on the application. The problem: users are upset they can't use the application without internet access. First, putting yourself in the shoes of the developers, think:

Is the application poorly designed?

What are my motivations and roadblocks?

Why does the application work the way it does?

Is there a technical reason a user can't work offline?

Think about what it will take from the developer's perspective to fix the user's problem. You might need to investigate what the issues are since you might not be a developer or CTO and have that experience. But it's vitally important to look at the problem as a developer. In our scenario, we will say that the application offloads the text overlay for photos onto the server rather than on the user's mobile device. This offloading is the reason that internet access is required.

So, let's take that problem to the next actor, the development manager. Your perspective needs to shift yet again. From the manager's perspective, the issue isn't just that a critical feature is performed on the server versus on the user's mobile device. That's a symptom to the manager, who knows how to fix the issue: move the functionality from the servers to the phones. Now, the problem for the manager is that moving this functionality costs engineering time—likely a hot commodity at your company—to make the shift. So, to the development manager, the root problem is the need for money and resources.

We aren't going to go any deeper or focus on the other actors at this point. But hopefully, this exercise helps. To get to a truly creative solution for any large problem, you need to understand the problem holistically. If you walk through all three steps, you'll have figured out the deepest root problem for each actor, and, from there, hopefully, you'll be able to start to understand possible solutions. Solving the issue may require more than one solution, but you'll at least know where to start.

With a complete understanding, creative solutions often come forward. So many people stop at the initial symptom and work to fix that. The underlying problem never gets solved.

Another common issue that I see is that people will stop after the first actor in the scenario. They find and (try to) fix their problem, but they never understand the full scope of the issues. For our scenario that may mean continually trying to put in application changes to fix the internet requirement but not understanding why users still won't download the application more. They don't understand that the problem runs deeper than that one application change.

When you're willing to slip into someone else's shoes, you can create an entire world in your mind to understand any problem from every angle imaginable. You are no longer constrained to your individual perspective, but you can incorporate the views of all stakeholders. When you can master this visualization, it becomes a game-changer. I have instinctively operated this way since I can remember. I'm not sure when I started, but I can tell you that I attribute much of my success to it. Not only does it help you grasp the whole scope of a problem, but it also teaches you empathy.

Sometimes I catch flack for this process. I've been told, "I don't need to know what my developers [or sales team, marketers, etc.] are thinking." Really? This product (or service) is your company.

Trust me, you want to understand their perspective. I encourage you to look at problems from every perspective that you can think of. Provide a more educated and creative answer. Practice this three-step process every day, all day—ABC.

4. BURN DOWN THE FOREST

One of the byproducts of the technology tsunami is that you'll need to let go of knowledge—of information around a particular technology because there will be a replacement. Let me demonstrate with a personal experience.

In early 2000, I became aware of VMware virtualization, which allows you to run many computers on one physical server. I saw this as a game-changer. It was very early and, at the time, not as useful as it would become later. Frankly, processing and memory weren't as advanced and cheap as they needed to be for mass adoption.

But I saw the potential, so I invested my time in learning the technology as much as possible so that I could become an expert. I gained numerous certifications in it, and I worked with the second-largest VMware virtual desktop deployment in the world. After ten years in the space, by 2012, it was becoming evident that the cloud was going to be the next big technology, which would cannibalize the tech that VMware (and I) had invested in. In particular, it was clear to me that Amazon Web Services (AWS) would be the big brand to usher in this technology.

Here, most people would get scared, and it was tempting for me to feel this way. But fear becomes a massive mental block. Once you've invested an enormous amount of time, blood, sweat, and tears into learning something, it's hard to let it go. It doesn't only apply to software, but to any skill, process, method, or technology that you've invested in. But, the tsunami *is* coming, and you must adapt. You can't stay put and "hunker down," hoping that you can hold your breath long enough to survive. You must "give up" the knowledge you've gained and go down the new deer path, allowing yourself to get all scraped up.

I started out small, tinkering on my own time with the available cloud tech, such as Azure, AWS, and, at the time, Rackspace. I was looking for indicators as to which one I felt would be the leader. After experimenting, I determined that AWS was the easiest to figure out and had the best user experience. I focused my educational efforts on that solution going forward. At the end of the day, I was looking for a technology that was going to be huge that could drive my career forward. I knew that the right platform would need to be easy to get started with and use, or it wouldn't be the leader in the long run. My assessment was correct, as AWS became the clear cloud winner.

Why did I spend all those years, over a decade, learning about VMware just to throw all the knowledge away? Truthfully, after pivoting, I would probably never touch VMware technology again. I knew it would be considered largely obsolete, displaced by the better, newer technology, the cloud.

But I didn't throw all the knowledge away. You can never really flush gained experience. Your brain compartmentalizes information. You have all of these synapses that get chained together to create understanding. The more those pathways are used, the stronger they become. When you don't use them, they do get a little weaker. But no knowledge stands alone; so, when you go down a new path, new synapses get linked together, taking your old experiences and breathing new life into them. I like to explain this with an analogy called "burning down the forest."

Forest fires are immediately thought of as bad things. But as it turns out, they provide much good. When a forest gets burned down, two things happen: First, the most robust trees survive and continue to grow, and, second, the fire cleanses the forest, as dead branches, along with weaker trees and shrubs, get burned away. This cleansing allows for new and stronger growth. What was once ground choked without sunlight now becomes a vast area available for new growth.

In your life, you need forest fires. Get rid of the information you don't need to move forward, and allow the critical knowledge to flourish. After the fire, you're never left with nothing.

5: KNOW THE QUESTION, NOT THE ANSWER

The last mental tenet we will discuss focuses on knowledge that is no longer valuable. When you went to school, the whole purpose was to drill you with facts that needed to be memorized. Then, knowledge was power. You had to memorize your multiplication tables. You had to learn about the Constitution and the Bill of Rights. You had to *remember* all of this information. In today's technological world, that's not the case. Knowledge is easily accessible. Everybody has it on their phone. If I want to know what the Bill of Rights is all about, I can Google that on my phone in about ten seconds and have every possible answer, and not just the ones that I was taught to memorize in school. *Knowledge* is no longer powerful. Even having the answer is not power. Knowing what questions to ask is power. The only place where knowledge is potent today is in knowledge that is so obscure that only a few people know it. But thankfully for the masses, that's only the case in limited circumstances.

Unfortunately, schools are still teaching memorization and that knowledge is power. My kids are in high school, and they're still learning how to memorize information.

Have you ever Googled something and not gotten an answer? Whenever it happens, it's infuriating. When it happens to me, it's almost always because I didn't know the right question to ask. Once I tap into the right question, I unlock the knowledge. That's power. You need to acquire the *background* knowledge, the crucial high-level knowledge that allows you to get to the answers. For instance, say I'm working on my Jeep, and I need to determine what part I need. I ran into this scenario recently when my daughter and I were restoring a Jeep Wrangler. There was a part of the engine we needed to analyze, but I had no idea what I was even looking at. Every search term I put in online couldn't get me the answer. It wasn't until I dug into enough of the background information that I was able to start asking the right questions. Once I did a little research on the parts of the engine, I figured out that the part in question was a small piece inside the gas vapor recovery system. The key, in that situation and to modern life in general, is knowing certain aspects of fundamental information, so you can transform that knowledge into the right questions. You don't need to focus on mundane details; you need to focus on the details that get you what you want.

When working on the Wrangler, I only needed to know, broadly, that the mystery piece was part of the larger vapor system. I didn't need to know that the piece is called a canister purge valve, a fact that I had to look up again just to write this sentence, proving my point: mundane details no longer matter. The best part is that not needing to know these facts allows your brain to store more relevant information.

Unfortunately, you must undo the conditioning that decades of education have drilled into your head. No longer is the minutia valuable. You're looking for the significant components that allow you to construct a mental image of the question that you need to ask. Those are the pieces of information that you need to retain and restore. Everything else is just noise, and it's going to get worse. The closer the technology tsunami gets, the faster information will be changing. Eventually, it will be moving so quickly that remembering any details about minutia is going to be completely irrelevant and a waste of time.

WRAPPING UP

We covered a lot in this chapter. The five tenets of tomorrow's surfer form the base of your (and your team's or company's) new mindset. First, you need to think like a child and look at things as if it's the first time encountering them in your life. Remember that facts are transient, and there will be more information in the future. Secondly, continue to prove yourself wrong to keep up with the changing environment. The third thing you must remember—and one of my favorites—is that you must always be creating (ABC). The successful companies of tomorrow will set the path, not follow it. Fourthly, be prepared to flick a match and burn down your forest to allow for new growth. Finally, as information becomes even more accessible, work on your ability to construct the right questions and don't focus on remembering answers.

You will need to practice these tenets. You will not read through this once and change your thought processes. It will take time to build these new mental muscles. I encourage you to exercise them daily until they become second nature. Once you have them down, it will be easier for you as a leader to demonstrate them to your team.

Now that you have a grasp of the mental tools required, let's move on to the next chapter and talk about the mental blockers working against your progress.

CHAPTER 5

How to Deal with Mental Blockers

Now that you understand the mindset required to succeed in the new technology world, we need to discuss the mental blockers that will hold you back. We will discuss several factors, but the largest factor, and the one we are going to talk about first, is risk.

Risk manifests itself in a multitude of ways. So we will explore in detail how this might look and how to overcome its adverse mental effects. Most often, risk introduces unnecessary delays— delays that you cannot afford if you want to maintain maximum company agility. After we discuss risk, we will move on to other important but less influential factors such as external pressures, compensation turmoil, previous experiences, and lack of trust. All of these items will create drag on your surfboard and prevent you from riding on top of the tsunami.

WHAT IS RISK, REALLY?

Risk, by definition, is a potential of danger, harm, or loss. It may not sound that scary when we say it like that. Or maybe it does. Either way, risk isn't really that simple.

Risk plays off complex emotions that are deeply ingrained in our brains. Identifying and evaluating risk is tightly linked to our fight-or-flight response. We are blessed with automatic responses that have allowed the human species to survive. We either defend against a danger ("fight") or we run from it (flight). This response served us well during the first couple hundred thousand years of human existence. Unfortunately, today in our civilized world, it's not all that useful.

Risk today is very different than it has been traditionally. Most problems today are not life-or-death situations. However, our brains tend to overreact as if every situation really were life or death. We get all wound up about our choices. I've seen this multiple times working with companies who are considering migrating to the cloud. After observing the agonizing, back-and-forth decision-making, I would think, *I'm not Rumpelstiltskin asking for your firstborn child. It's just the adoption of new technology.* Unfortunately, fear rears its ugly head too often in business. There's uncertainty around every corner, and we allow fear to block our progress. It's nothing to be ashamed of. Everybody has the same responses—actually, not everybody. There is a group of people that doesn't feel fear—psychopaths. Nobody signs up to be in that group. The rest of us must learn how to compensate for the natural tendency to fear the unknown.

If you want to outrun the tsunami, you will need to get past the mental blockers of risk. If you don't, you will not survive. You cannot afford analysis paralysis. I'm not saying to throw caution to the wind, but don't get hung up on the emotional part. Think about the problem logically and proceed forward.

ACTION IS THE OBVIOUS CHOICE

In today's rapidly changing and accelerating world, you can't afford not to take action. Lack of action means drowning. While you're worried about some *potential* consequence of taking action, a competitor is *actually* charging forward.

Today we live in a global economy without the same geographical boundaries of business from the previous decades. You aren't only competing with firms in the next town or state. Now, you must take action like the life of your business depends on it. Because it does. You're competing with smaller companies, whose founders' *literal* lives may depend on *their* success. So, their mental barriers are entirely different from yours.

You may be fighting for a business, but someone may be fighting for physical well-being, and when someone's fighting to survive, their brain tells them they must succeed. For some, the alternative to beating you is to go hungry or live in a crime-ridden area. Maybe it's not always that drastic, but perhaps your competitor's founders have invested their life savings into their business and need to succeed. Either way, their mind is 100 percent in fight mode. These people are going to be coming for you with everything they have.

If you're reading this book, I'm guessing that you're not in that position. You work at a decent company, or you are doing pretty well for yourself independently. When it comes to risk, your brain isn't leaning towards fighting for your life. It's leaning towards *flight*. You are afraid of what might happen because you do have something to lose. This stance will always put you at a disadvantage. Plus, with the proliferation of cloud computing, a company anywhere in the world can put infrastructure right in your backyard, providing quality service all over the globe. Providing this form of global infrastructure in the past would have been a difficult feat and was a barrier to competition.

In addition to the global competition, startups will continue to threaten any established company. However, the tech tsunami is going to increase the pressure applied. The ease of innovation and the ability to test new technologies and products quickly will increase the number of startups over the next decade. You will see more and more startups emerge because the barrier to entry is so much lower than it was previously. You can start a new mobile app very cost-effectively today. Using the right serverless technology, you can launch that application with next to zero infrastructure cost. It no longer takes hundreds of thousands of dollars to purchase equipment.

These startups also look at risk differently. They either have very little to lose, as in a pilot experiment to see if there is traction, or they have everything to lose due to massive personal investment. Either way, they will see inaction as the risk, not action.

Money follows action. In the last chapter, we already discussed how society has turned into a giant game of instant gratification. When it comes to risk, if you don't take action, someone will run away with your lunch. You cannot let your evaluation of risk stop you from moving. Action is always the obvious choice.

Some enterprise companies out there say they're nimble like

startups. Don't fool yourself. Even Amazon, a company that touts itself as an agile company, is nowhere near as nimble as a startup. Amazon has a process called a six-page narrative. If you want to get something done at Amazon, you must write one of these narratives to get the funding and approval. Sound easy? Not even close.

What isn't talked about is the 100 pages of appendices (with charts, diagrams, market data, and models) or dozens of revisions that someone writes before presenting their "six-page" narrative. If a startup were to act that way, they would be out of money before they got to the tenth revision. If *Amazon* can't deal with risk swiftly, what hope does any other enterprise have? A startup is a startup only while a startup. Once they get bigger, they must stop pretending like they're a small, nimble outfit. Instead, they need to shift their focus to reducing team size—not by firing people but segmenting the work that they're doing into more agile units. With increased engagement, better communications, and less layers of management, smaller teams collaborate better. Seek to create mini-startups that operate under your larger umbrella and have the autonomy to get their jobs done efficiently.

With the speed of technology, fear is going to increase. You won't have as much information available to you as you would like to assess risk and make a decision. Information now changes so quickly that it is impossible to make a "perfect decision." You must take action to ensure solvency.

LIFE IS A TWO-WAY DOOR

When it comes to risk management, most fear is unfounded. You worry about something happening, and then it never does.

Let's say that your teenager just got their license. You are worried about them and the "high" risk of them getting into an accident due to being inexperienced. Life, being the cosmic comedian that it is, doesn't put your kid into an accident, but instead, some neighborhood hoodlums take your car for a joy ride and then torch it in the local parking lot.

Where did all that worry get you? Nowhere. Something you could never have anticipated happened. The same holds for your company. When it comes to technology, your mind will construct all sorts of fears around adoption. And the truth of the matter is that most of them will never come true. The one thing I can guarantee will come true is that if you wait too long to take action, your competition will take your business. And, likely, you won't lose to whom you think your existing, established competition is. Most of those companies are probably also overanalyzing and overvaluing risk. They're struggling to make decisions as fast as they should and assess risk more quickly and accurately. The new startups are the real threat, and they don't have those same fears. They're charging forward, and they're going to damage your place in the market significantly if you stand idly by. My suggestion? Address risks from the standpoint that life is a two-way door.

This two-way door is a concept that I learned while working at Amazon. I like it. So I'm stealing it. The idea of the two-way door is that you shouldn't walk through a door that you can't walk back out of. There should always be an escape plan. Even though your brain will be telling you everything is a one-way door, you'll see that most of them are two-way, and you can easily undo most decisions. Let's look at a scenario.

Say you're thinking about creating a new mobile application.

You're evaluating the risks of making this application. There's financial risk, the cost of developing the application. There's the reputational risk of people not liking the application and your company losing some traction in the marketplace. There's operational risk—what if it isn't designed correctly and it goes down and thus creates more reputational risk? Finally, there's a security risk because you may inadvertently share data through a breach, poor design, or poor programming.

Most of these things are reversible. You could abandon the application due to poor traction. You might not think that is an option. However, Google has tried and scrapped so many different products that they're great at it. Google is good at falling fast. Has it hurt Google's reputation to do this? Maybe a little, as some of us now have a slight amount of fear that if we use a Google product, they might suddenly kill the project. But in the grand scheme, Google has a pretty good reputation for delivering reliable, solid products that last.

Let's consider the risk of a security breach, which would also create reputational risk and likely temporarily affect stock price. There have been quite a few significant breaches in recent history (Target, Equifax, Home Depot, TJ Maxx), and they are all but forgotten. Plus, breaches are becoming more common as time goes on. You don't want to intentionally misuse security, but it's important to note that the long-term impact of a breach is less today than it was in the past.

Reputational risk can be a one- or two-way door, depending on the extent. In the case of Arthur Anderson, their accounting scandal was a one-way door that sank the company. But there have been plenty of other issues that have only affected a company's reputation for a short period. Case in point: do you remember the Facebook data breach of 2018? Has that stopped you from using Facebook today? As I said, most reputational risk is a two-way door. You can always get back on track with the right marketing and customer obsession.

I also consider financial risk to be a two-way door. Unless you totally bankrupt your company, which would be pretty hard to do while developing an app (again, I'm assuming you're at a decently well-established firm), you can always cut costs and work on increasing top-line revenue to recover. It might be difficult, but it's doable.

There are a few things that are actually one-way doors, but they typically have to do with pricing. If you charge for a product, you can always increase the price. But if you decide to launch a product for free and then turn around later and want to charge for it, you may have a hard time overcoming that. It's hard to get people to start paying for something they were getting for free. So, it's essential to nail down your business model to avoid that one-way door.

Keep in mind that when you're working in this new hyperspeed world, you've got to categorize risk into one-way- or two-way-door paths. Pay more attention to those one-way doors, and don't get hung up on the two-way doors; evaluate and move on.

ACCEPT, DON'T ELIMINATE

One of the biggest mistakes people make regarding risk is that they think they need to eliminate it. That is absolutely *not* the case.

When I worked in IT security, many security managers tried to eliminate risk. They thought that we should try to obliterate risk. But that's impossible. I take that back. It's not impossible. But to do it, you can't do anything. So how do we handle this dilemma? Well, risk is something that should of course be mitigated, but also accepted. That's the keyword, *accepted*. You don't need to obliterate all risk. Analyzing risk is a lot like making maple syrup. All the tree sap goes into a big bin, and you keep boiling it down until you get maple syrup. That's the same thing with risk. You take the risk of doing something and put all that risk in a bucket, and you keep boiling it down. Work on mitigating it away until you're left with the residual risk. That residual risk is your maple syrup, although not as tasty.

Risk is inherent in everything that we do, right? Every time you drive to the grocery store, you're taking on risk. You're taking a chance that that person who's driving forty-five miles per hour directly towards you is going to stay in their lane, not swerve over and strike you dead. You have accepted that risk, which is a residual risk of driving. It is what it is, and you're not going to get away from it. But for some reason, when people get into business, or they run a company, a department, or a team, they want to eliminate it. You can't; it's impossible. So let's not try.

EXTERNAL PRESSURES

Whenever you set out to do something in life that isn't an accepted norm, you're going to feel pressure from outside forces. This pressure isn't just related to technology either. It applies to life in general. You're going to have many naysayers telling you that something won't work, or it's going to be too expensive, or insert whatever detractor they have. You'll need to get used to it. That's how things are going to work in the future. You've probably already experienced this, but it's going to get worse. It's going to happen more frequently as technological advances accelerate. You must establish that getting negative feedback is OK. There are plenty of self-help books out there that detail how to address and compensate for external pressure. I want to highlight that, when it comes to business, external pressure only comes if you do something new. So, external pressure is an excellent indicator for you as to whether you're keeping up with the tsunami or not. If they're not saying you're crazy, then you probably aren't pushing the envelope far enough. Somebody should think you're crazy because that's where all the best ideas come from.

COMPENSATION TURMOIL

Another area where you'll probably find a mental blocker is when it comes to your compensation. Today, most people's earnings are somehow tied to a bonus linked to corporate, departmental, and individual benchmarks. Your compensation might also be connected to expenditures or sales. In the rapidly progressing future, there's going to be an initial disturbance to your bottom line. There may be significant expenses from branching out with new technology, or you may experience a short-term loss of sales due to the cannibalization of existing products or services. For instance, assume you're running a desktop-based application today, and you want to transition to a mobile application. You're potentially cannibalizing your desktop users to make them mobile users. You also have an expense for creating a new version of the application. It's also highly probable that you can't charge as much for mobile applications. Ultimately, you'll need to understand that you're looking at the long-term goal, and there might be some short-term compensation penalties.

I've seen this numerous times where a mobile application will be significantly cheaper than its desktop counterpart. People are not willing to pay $500 for a version of Photoshop for an iPad, as they may for a desktop version. Photoshop for the iPad costs *$80 a year*, which reduces revenue by 84 percent. Are you mentally prepared to see an 84 percent reduction as you expand your user base?

In this scenario (and others like it), you're looking at increased expenses in the short term (for the application development), and potentially decreased revenue, not to mention an extended lead time to grow a customer base. There's a high probability that this will affect your year-end bonus, which is a big inflection point, because that involves your family—potentially your spouse, kids, dogs, and even the parakeet. The short-term effect on your compensation will significantly impact your decision-making. This reduction will be a blocker. You'll need to move past that blocker to understand that technology adoption will net you more compensation in the long run. If you fail to move forward, your compensation will be affected in the long run, whether you like it or not. Market forces and the reduction of market share will do this for you. Moving forward now is up to you, and I would rather control my own destiny than let others do it for me.

LACK OF TRUST

Whenever a new technology comes out, I'm inherently skeptical. I blame marketing. When a new technology is released, there are many promises. However, most times, the technology isn't completely baked yet. It's still a bit gooey in the insides, and I've seen my fair share of projects get gummed up upon delivery.

So, for me, I have an inherent lack of trust, which means, overall, this is one of my ongoing mental blockers, and it's probably the hardest one for me to work through. Unfortunately, there's been plenty of technology that's supposed to deliver something, and then at the end of the day, it can't really do what it said it could do. I observed this all the time in the security space. A system would tout that it's the next greatest thing, stopping hackers in their tracks and catching every virus that's ever been written. Unfortunately, in the security world, once there's a new technology out there, attackers change the way they attack, rendering said technology less effective and, in some cases, completely worthless.

Because of this, I now have an inherent distrust of new security technology. I tend to wait and see. For me, this is a mental blocker—because the competition creeps up on me while I'm waiting and seeing.

Here's the ugly truth about the tsunami: The ruse put on by vendors is going to escalate. With innovation moving so quickly, to remain competitive many vendors are going to flaunt features and capabilities that don't really exist. It's going to take some mental fortitude to continue to work around the mental barrier of lack of trust when you keep getting burned by broken promises.

My solution to this problem is to compartmentalize the issue. I must remind myself that just because one vendor didn't live up to expectations doesn't mean that another won't. I don't want to become a jaded ex-lover that's bitter, lonely, and stuck not moving forward. Additionally, I can practice the saying of "trust, but verify." There are ways to balance the memory of unmet expectations with moving forward. A few safety measures include contractual back-out clauses or spread-out payments.

Lack of trust isn't limited to just computer software and hardware. You might be too young to remember this one, but Ford put out a car called the Pinto. The Pinto had a nifty feature—if you got rear-ended by someone going faster than thirty-five miles per hour, your Pinto would catch fire. I imagine that for anyone who's purchased a Ford Pinto and been rear-ended at high speed, there's a sincere lack of trust in Ford. The carmaker destroyed the trust of those buyers.

PREVIOUS EXPERIENCES

The last mental blocker that we will cover relates to previous experiences. Although the trust that we just discussed is a byproduct of prior experience, that's not the angle we are looking at here. Previous experiences can affect you in two different ways when adopting technology. The first, and typically the most problematic, is that previous experience causes you to drift toward technologies you've used before. It's similar to the deer path analogy. You will feel more comfortable with technologies you've used before, even when there's a possibility that a newer technology would be a better fit.

Humans tend to drift to the known. You must work against this tendency. If you're running into this mental blocker often, try working through some of the techniques we discussed in the last chapter to ensure you're investigating all potential avenues.

The second problem with previous experiences occurs when you've had a terrible experience. When you work with technology and have a poor outcome, you're unlikely to ever use that technology again. At first glance, this would seem to make sense. After all, what would make it work this time if it didn't work last time?

But you must steer away from that thinking. Technology is very complex. The circumstances that caused the technology to fail in your prior experience don't predict future failure. It's probable that using this failed technology now might not yield the same results.

If you have tried a technology a couple of times and never had a good result, you wouldn't want to try again. But at least give it a second try.

Also, remember: technology moves fast, and, depending on the time between your last failure and the present, your previous experience might not be valid at all. New technologies often advance fastest when they are very fresh. The more infantile the technology is, the more invalid your experience. I've experienced this firsthand with the cloud. When I first tried out the cloud, it was buggy and had many outages. It wasn't ready for prime time, so I abandoned it. If I would have kept that attitude, I wouldn't be where I am today, having seen the amazing technology advances that the cloud offers. You must reevaluate your stance as technology advances.

WRAPPING UP

Getting out of your own way is a constant battle in life, not just with technology adoption. The best results are gained from mental fortitude and working past the artificial barriers created in your mind. In your business, you must be the leader who can deal with previous issues (such as untrustworthy vendors). You've got to keep an open perspective. It's your responsibility to understand how your sales team will be impacted by changes in the business and how they will get paid if you want to go in new directions. When the fear of risk rears its ugly head, attempting to stall you, remember: (1) You can back out of most technology decisions, and (2) money follows action.

Now that you have a handle on the mental blockers that you could experience, you'll be able to assist your team better to overcome these hurdles. Everything you're thinking and feeling will be similar to what they're feeling. Remember that, and empathize. Often, companies fail to understand and address how their staff is thinking and feeling during technological change. Being cognizant of their emotional state will put you further ahead than your competitors.

CHAPTER 6

Waxing Your Surfboard I

Culture Changes

Now that you have a strong understanding of what you need to do to be mentally prepared for the impending tsunami, focus on what you need to do to keep your company from being disrupted.

To get ready to surf the tsunami, you must wax your board to ensure you don't slip off once you get moving. That's what you need to do for your company: ensure that it stays sure-footed and maintains traction in a fast-paced environment.

There's a lot of ground to cover when it comes to preparing your company for technological advancement, so I've split "waxing" into two chapters: we'll cover culture changes in this chapter, and focus on process changes in the next.

The significant change you must ultimately observe in your staff is the necessary mentality shift discussed in Chapter 4. Plus, they must be able to overcome the mental blockers discussed in the last chapter. Thinking clearly is vital to your company's long-term success, but you can't stop there. You must cultivate the right *kind* of people within your organization. And, you must ensure they're developing the right work habits.

When I talk about people and habits, I'm talking about changing company culture. Unfortunately, company cultural changes are the hardest kind of changes to make. In fact, culture will be your biggest hurdle in preparing for the tech tsunami. Changing it will take much time and even more effort.

The culture of your company is established by the people who work there. Unfortunately, we might be talking about some challenging subjects. Sometimes people won't just fit with the new narratives you want to write.

What's the desired company culture? In the new, hyperspeed world of technology, you will need a nimble organization, and being nimble isn't just based on *processes*. Nimbleness is based on how your employees think and work. They must not only be flexible, but they must be able to adapt quickly. Let's start with the essential function of your new culture, creativity.

CREATIVITY

Creativity is going to be the new corporate currency, the "bitcoin" of your company's future. Being able to see the world in different ways will drive your future success. After all, you don't want to be the next Sears or Blockbuster. Creativity is key, because, if you're being as creative as possible, you are, in effect, working as a startup. If you're established, you can't actually *be* a startup. As we discussed earlier, established companies have defined processes that prevent them from truly being startups. Of course, being creative *like* a startup is only the first step. It would be best to execute like a startup, but we will cover that in the next chapter.

Creativity is vital, because it's the grease inside of your company that allows it to get off the deer path. (In Chapter 3, we covered why it's so hard to get off the deer path. Essentially, your brain doesn't want you to try anything new, because it feels comfortable when you're doing what you've always done; that goes for the individual and the collective "brain" of an enterprise.) Creativity allows your employees to take the blinders off, look at things differently, and see new potential paths.

There are dozens or even hundreds of ways to use technology in your company. It all depends on what your organization does and what industry you are in. You could be a railroad that invests in Internet of Things (IoT) technology to monitor rail cars, an airline that invests in AI for route planning, or a manufacturer that purchases machining equipment. We couldn't possibly cover enough to include every scenario. So, to make this an effective discussion, we will cover technology from the perspective of an IT department. Most companies, other than tiny ones, have an IT department.

The biggest problem with IT is that it's an afterthought within most companies. Information technology started as a function of finance. Companies used computers for the primary purpose of running the accounting systems within organizations. Chief Information Officers (CIOs) predominantly answered to Chief Financial Officers (CFOs). Times have changed. Now, CIOs answer to the CEO, which streamlines things, but still, unfortunately, most view the IT department as a cost center.

Historically, IT has cost the company money. It didn't generate revenue. Therefore, today, it still never gets the full funding that it should. But does your IT department *have* to be a cost center? If you look at it more creatively, is there something within IT that you can monetize? Perhaps you could monetize your internal applications for use externally. Several companies do this—they've pushed their IT departments from conventional cost centers to profit centers. You'd be surprised at all the ways you could do this. And, an avenue doesn't need to be in line with your company's core offerings, either. Several industries have had great success with this model, including airlines (who resell their software), banking (such as Capital One Shopping), and insurance companies (who often sell valuable data). Ninety percent of companies have never considered such a creative avenue.

I consider myself creative. I've been interested in music and painting my whole life, and I even wanted to be a voice actor. As a creative, I view technology as my canvas and paints as technologies. When I need to solve a business problem, I envision a canvas. Why? Well, because I need to envision a blank slate. With a blank canvas, you always start with what can be, not what is. You build artificial constraints around your solution whenever you start with what exists. Now that I have my empty canvas, I bring in my palette of technologies that could solve my problem. For each issue and industry, the pallet will be different. (If you're in oil and gas, the technology available will drastically differ from what would be available to you if you were building skyscrapers.)

For instance, your available paints might include blockchain, CNC machines, the cloud, or LIDAR. Individually, these are just paints; they don't have much value until you put them together to create a masterpiece. Likewise with your creativity: When you and your team take an available palette of technologies, and sprinkle on some creativity and start applying them in a strategic way, you start to create value, just how an artist does with her paints.

Here's how this works in practice:

Cars are a great example of painting with technology. Car companies spend millions of dollars coming up with slick new concept car designs. These cars are made of thousands of pieces that are built using all sorts of technology. Some pieces are pressed steel, made in giant hydraulic presses. Other parts are made with CNC machines. Most of the interiors are made using injection molded plastic. Like we discussed before, some brake calipers are now being 3D printed. The designers at the manufacturers use these technologies to paint a new car.

When you use this process, remember: be comfortable with starting with a blank canvas. You don't want incremental changes from what you had before. You want a full, new Picasso.

HIRE FOR CREATIVITY

Overall, you want employees who can look at the market and see something radically different than what anybody else sees. Those are the team players who are going to take your company forward, who will keep you out in *front* of the tsunami. Those are the critical minds who can make your business successful. Oh, and you can't just rely on one. You're going to need armies of people in every facet of your business in this new creative model, with everyone looking at problems differently. While an owner or CEO sets the precedent as to the skills and mentality to hire, it's up to everyone at a company to look for these traits. Why should everyone have this responsibility? Because almost everyone in a company interviews potential hires, not just the owner; when was the last time you were in an interview and only spoke to a manager? Usually, a manager and a team member will interview you. Those team members should be looking for the same traits as the manager or owner.

Cultivate team members by teaching them how to think and what to look for. Doing so will propel your company further ahead of the competition.

As a reminder: If anyone ever comes to you and says, "That's the way we've always done it," I encourage you to fire that person immediately. If someone ever uses that sentence, I can say with absolute, 100 percent clarity that your company would be better off without them. They are not only toxic to the culture that you're trying to create, but their thinking is also contagious. It will take them less effort to undo your work than it will take for you to do it.

TOXIC EMPLOYEES

You want to cultivate your most creative team members, but you also need to remove toxic employees.

Typically, when we're talking about toxic employees, we are talking about, to put it simply, jerks. But when we're talking about innovation and technology, we aren't talking simply about jerks; we're talking about people who are stuck in their ways.

Whenever I find someone like this, I like to think of them as a

"George Jetson." From the old cartoon *The Jetsons*. (I may be dating myself a bit here!) The show was somewhat like *The Flintstones,* but rather than being set in the prehistoric past, it was set in the far future, where everything is basically push buttons and automation. George Jetson worked at Spacely Space Sprockets, and in his day job, he sat in his chair and pressed a button. George pressed that button day in and day out every day. Even though computers and robots ran everything, people still needed something to do for work.

I assure you there are plenty of George Jetsons working in organizations right now. They may not press the same button all day long, but they're so used to doing what they do, that they will be an almost immovable force to get them out of that rut. These are people who actively hate change. They hate change so much that they will attempt to derail projects that force them to change. It's a bit ironic: I've seen some of these people put in more effort subverting change than it would have taken to simply change.

With the incoming technological tsunami, you don't need people putting cement shoes on the rest of your staff. You can't afford the drag a George Jetson will put on your organization. No matter where these people work—in the boardroom or the mailroom—they will sink your organization. (However, the higher up the ladder a George Jetson sits, the worse their impact will be.)

For the last seven years, I've been doing significant migrations to the cloud for Fortune 100 companies. I can say with certainty that if you have somebody in the C-suite who wants to derail an innovation project, they will. For instance, a director of infrastructure at a company looking at moving to AWS didn't want to give up control to the cloud. However, when assessing the return on investment for moving to the cloud, it was clear that AWS was significantly cheaper than continuing to use their on-premise hardware. We'll call this director "Larry." Larry purposely left out certain costs in his analysis to support his narrative, thus tipping the scales to his preferred path of staying with the on-premise infrastructure. Trust me, these kinds of people are incredibly toxic to your organization's forward momentum.

Some leaders may simply say no. Others have more pernicious intentions, and they may outright lie or stack the deck against your initiatives. Owners: be extremely careful about whom you appoint to positions of power.

MAKE IT WORK VS. SEE IT WORK

There are two types of people in organizations. When given a task, some people will always "make it work." These people don't get hung up on roadblocks; they figure out a way around them, working tirelessly to make everything that needs to happen, happen.

This level of effort is dramatically opposed to the work ethic of people in the "see it work" camp. The "see it work" crowd are the people who—the second they find a roadblock, any reason why something won't work—throw up their hands saying, "Oh well. I tried. It didn't work."

Let's look at this scenario. Rob is a "see it work" person. He's been asked by his manager to look at a new accounting system. The boss wants to know if there's something a bit more modern, such as a SaaS or internet-based technology, instead of their current installed systems. Overall, the boss wants something that is a bit more user-friendly. Rob's company has been using a somewhat proprietary accounting system, built for companies in the machinery maintenance space (fixing forklifts, semi-trucks, and excavators).

Rob investigates the various offerings. He identifies seven pieces of software. He creates a pros and cons list for each one and identifies any non-starters (items that will prevent his company from using any particular piece of software at all).

Rob analyzes each option, to see if the software can import data from his company's existing systems. He finds that six out of the seven vendors cannot import the current data. Rob immediately disqualifies six out of the seven software packages. The seventh package, even though it can import the data, charges ten times as much. The fee is far greater than what Rob's company is paying today. So, Rob disqualifies the seventh and final package. He delivers his findings to his manager, who reviews them and agrees with Rob's assessment. The company stays with the existing software.

Let's contrast that to Tim, a "make it work" personality. For simplicity, let's say Tim works for the same boss, at the same company, with the same parameters. Tim also identifies seven different software options. Tim also determines the six packages do not accept an export of their existing data. Tim's somewhat discouraged, but his wife works in IT as a data scientist, so he wanted to get her opinion.

While at the dinner table, Tim asks his wife for her opinion, and if there is any way to convert the output from their existing systems into a format that one of the new systems could understand.

Tim's wife says, "Someone will need to do a lot of mapping to make that happen, particularly mapping fields in the old system into the respective fields for the new system. But it's doable." She explains that to make a better assessment, she'd need more information about how both the old and the new systems' databases are structured. Tim researches the data structures, and brings the database layouts to his wife, who does her analysis and determines that the table structures are very similar. They'd be able to build a translator in two to three weeks of engineering time, including testing and deployment. With that translator, they should be able to migrate data from the existing system to the new system.

Tim repeats the process for the other five softwares. Using this information, Tim completes his report. Tim's report shows that six out of the seven technologies are plausible. (He eliminated the seventh due to cost.) He then details how much engineering time it would take to create a translator to translate their existing export files into files for the new systems.

Tim's analysis shows that one software package would be the most cost-effective, while meeting all of their other requirements. He delivers the report to his boss, and after explaining the situation, he suggests they move forward with purchasing this particular software package.

As you can see, the see-it-work and the make-it-work mentalities lead to vastly different outcomes for your business. The see-it-work mentality is highly detrimental to your company's long-term success, as it limits the ability for changes to compensate for technological advancements.

Letting someone go is never fun. Unfortunately, if you want to succeed, you will need to eliminate the see-it-work crowd from your business. It's better to never hire those kinds of employees in the first place so you can prevent the toxicity from even entering the door. But that's not easy to accomplish. There is no straightforward question that you can ask someone to determine if they will give up on something the first time there's a roadblock. You will need to use behavioral questions and come to your conclusion about them in a roundabout way. For instance, you could ask something like, "Tell me about a time when you were asked to lead an implementation project." Or you could ask something like, "Tell me about a time when you were asked to improve a process." The answers to those questions should help you determine if someone gives up too soon for your liking. You're looking for someone who always finds solutions, even when they are repetitively blocked. This pattern demonstrates the make-it-work mentality.

CONTINUAL EVOLUTION

The next component of culture that we will talk about is making your company more adept at change through constant evolution. First, we must look at how companies typically adopt technology. Most companies will implement some technology—for instance, a software package—and then go into a hiatus of change, potentially for years. In that scenario, their general technology adoption trend looks like the jagged edge of a saw blade—up and down, up and down.

You want to create a culture of evolution where you're constantly changing versus making only giant leaps in technology. Instead of being that jagged sawtooth, you want to be smooth like a razor, a continuous stream of evolution.

Creating a culture of constant evolution is not something that your team or company can achieve overnight. It takes a substantial amount of time. What you're doing by making this transition is getting your people used to change—accustomed to being in a state of transition because it's always happening. In this state, there's always something new to learn, something new that requires a process change or the development of a better idea to achieve company goals.

Using the constant evolution method is significantly better for your organization in the long run. When you use the sawtooth method, there will be times where there isn't change and people get comfortable. You don't want your people to get comfortable. When they get comfortable, they resist change. Instead, by creating a small stream of changes, you can keep your people adaptable. Minor changes are also significantly easier to implement than large ones.

An additional bonus of using continual evolution is that your organization can get further, faster. Like in the tortoise and the hare story, the tortoise is slow and cumbersome, but consistent. That's the type of constant evolution we're looking for. Meanwhile, the hare starts, stops, starts, stops. And we all know who won at the end of the story (the tortoise). Slow and steady wins the race, especially when your competition uses the start-stop method. Every time they restart, they are standing in the equivalent of tar; no one wants to move. Everything is changing, and that's what you want your people to do—change. You want your people to roll with the punches. As the leader of your team, you set the pace and the path. The way may not always be perfectly straight, but, still, set a course. Perseverance wins out. People tend to think that being evolutionary isn't a good thing when it comes to technology.

People like to imagine huge leaps. They want to be revolutionary by doing something extraordinary. But as we discussed in Chapter 1, all the technology we have today is built upon thousands or tens of thousands of innovations. Technology *is* evolutionary.

The difference we see today, which is the whole premise of this book, is that slow and steady change that historically took millennia is happening faster and faster. At some point, change will be fast and steady, and that's what we need to prepare for. Right now, we're at the end of the slow and steady phase. We have a little time yet to prepare our organizations for the transition. When slow shifts to faster and too fast, if you've been practicing evolutionary change, you'll be prepared.

Small continual changes also prepare your staff for large changes, so staff will be more accepting of them. Even if you're working in an environment of continual change, there's going to be a point where a major system needs to be replaced. Continual evolution greases the hesitancy against those changes significantly.

GENERATIONAL CHALLENGES

One thing that I would like to highlight about a culture of continual change is that creating one will pose some generational challenges. It's not a hard and fast rule, but in general, the older you are, the less adept at change you will be. Obviously, that's not true for everyone. But in general, it's a trend. You can tie the speed of the tsunami directly to each generation. The older a generation, the slower the tsunami was building, so that's what people of that generation became accustomed to. My children have seen significantly more change than I have, and I've seen more than my parents.

Looking at how technology changed between 1950 and 2000, there was nowhere near the level of change we witnessed in the 2000s and the 2010s. What does this mean? It means new generations are adept at change because they're used to it. If you go young enough, like my daughter's generation, Generation Z, they accept change *daily*. Getting a request to install a new app on their phone happens every day or so. Many of the apps kids use today are gone tomorrow. They are off using some new tool to filter their TikTok videos. To this generation, most apps that they use are extremely transient.

I'm not telling you to go out and fire all your older people. For one, that's illegal. In addition, the older people in your organization have a wealth of knowledge. Older generations have different lenses, and they see problems differently. Their expertise and insight are invaluable. I am telling you to be aware that older generations aren't as accustomed to a high rate of change. Compensate within your plans to accommodate for that. Some employees might need some additional training. Some might need preparation and coaching.

WRAPPING UP

Changing culture isn't easy, but now you are equipped with the three key ingredients to keep your company successful when the tsunami makes landfall. You have applied your first layer of wax to your surfboard to ride the wave when it comes. But creating a culture of creativity and continual evaluation without toxic employees is only the first step. In the next chapter, we will cover the next layer of wax, helping you change vital processes to adopt new technologies.

CHAPTER 7

Waxing Your Surfboard II

Process Changes

Getting your culture prepared is only the first step in riding the wave. The next puzzle piece requires that you eliminate some archaic processes in your organization that don't fit with modern technology. The older and more established your company is, the more probable that you will have outdated processes on your team. Conversely, if your company is young and/or small, this chapter will not be as applicable to you, *today*. Nonetheless, it's essential to understand these concepts as you grow so that you can prevent blockers of technological adoption from growing within your institution.

PURCHASING

The first process we are going to look at is purchasing.

Talking about purchasing might sound peculiar when we're discussing technology, innovation, and the speed of change happening in the world. But unfortunately, many companies have purchasing departments, and the required processes of these departments often inhibit the adoption of technology.

To demonstrate why a purchasing department procuring technology is outdated, let's break down the purpose of the purchasing department. This department has one primary role—to buy goods and services for the company while maximizing savings, which sounds great. I mean, you do want to get the best price. Unfortunately, this department process inhibits your ability to adapt to technology quickly.

1. PURCHASING MAY SACRIFICE SERVICE OR QUALITY

First, purchasing departments will want multiple quotes when purchasing technology. This can be particularly problematic when you're engaging with consulting services. Frankly, it's tough to make an apples-to-apples comparison between two consulting firms. Some firms, like Deloitte, are geared towards building a long-standing relationship with a company and working as a strategic partner. These types of firms often charge more because they bring more value to the table. Other firms are what are referred to as "body shops." These firms simply apply people to get individual projects completed, using a more transactional approach, and they typically cost significantly less.

It's also difficult to compare two different technology providers, even if they're offering the same piece of technology. Likely, they'll have drastically different capabilities (AWS and Oracle Cloud are technically competitors but have vastly different capabilities).

Further, customer support options vary greatly between vendors. Your purchasing department will always try to get "three horses in the race" to get the best price. They want multiple vendors so they can drive competition between them. However, the company with the best price might not always have the best service or the best product.

When I first started my career, I worked for a small, white-box computer company, and we were a premium brand, certainly not the cheapest. However, we offered a quality product, and we stood by it. Plus, we had excellent customer service. While I worked there, customers would often say that our price was higher than our competition's. And my boss would say, "You can have price, quality, or service, but you can only pick two." I have found this to be 100 percent correct.

Service, support, and features are vital to your business's ability to ride the wave. You're probably not getting good quality if you demand the best price and great service. If you're getting a great price and excellent quality, you're probably not getting excellent service. While your purchasing department is hunting down the best price, they may be doing you an extreme disservice.

2. PURCHASING IS GEARED FOR COMMODITIES

The second way purchasing departments slow down adoption is that they're primarily designed for commodity-type items. Purchasing departments are great when buying printer paper or 1,000 sets of shelving. There are multiple providers, and they all provide relatively the same product. But this equality doesn't exist in technology, where a single provider could be the only firm offering a particular piece of technology.

A company that I was consulting for wanted to move to AWS. The purchasing department required competitive quotes. At the time, AWS was drastically ahead of the other cloud providers. There was no competitive quote to be had. Ultimately, this caused a bunch of delays, because the purchasing department didn't even know how to handle the situation. Management had to step in to get things moving again.

Even today, there's no way to crown one winner from the three major cloud vendors (AWS, Azure, and GCP). They all have their strengths and weaknesses. You can't possibly get "competitive quotes" and compare them, because that's not how the cloud operates. The major vendors don't offer all the same services, and even if they seem to at first, they have different feature sets once you look into the details (i.e., look past standard servers and disk). The only time a purchasing department can effectively operate that way with technology is when the technology has already been commoditized, which means you're likely not doing anything novel by purchasing it anyway. There are very few use cases in today's world where a commodity will give your company a competitive advantage.

You don't want to be a laggard company; you want to be at the forefront of technology. So, unfortunately, your purchasing process (for technology purchases) has to go. I'm not saying to get rid of the entire department for everything. There are still many savings that can be had when it comes to buying toilet paper for the bathroom, but this department has no business being involved with your technology purchases, plain and simple. If you want to win, it's got to go.

EFFICIENCY < EFFECTIVENESS

I'm about to throw a couple of "e" words around so let's level set on what they mean. Efficiency and effectiveness—these terms sound confusingly similar. Commonly used in medical research, project management, and decision science, they are often mixed up in everyday conversations. If you're in a hurry, here's the difference:

- Efficiency means working as intended, quickly and without wasted effort.
- Effectiveness means working *on the right (or best)* tasks and strategies.

Efficiency is all the rage when it comes to technology, and many people get hung up on efficiency. They talk about efficiency like it's cocaine. Working efficiently means doing things with the least amount of effort. For instance, using an excavator to dig a hole instead of using a shovel is a demonstration of efficiency. The act of digging a hole hasn't changed, just the mode in which you do it. The excavator saves you time and effort. You want to be efficient. But what's most important of the two "E's" is *effectiveness*.

To demonstrate why effectiveness is vastly superior, I like to talk about Formula One and pit stops. As the race cars go around the track, they burn up tires and fuel; throughout the race they've got to come back to the pits to refuel and change those tires.

In the 1950s, pit crews in Formula One were very efficient. They could refuel cars, polish the windshield, and change out tires in sixty-seven seconds. I don't know about you, but I'm pretty impressed with sixty-seven seconds. I couldn't dream of changing the tires on my car that fast. They were very good at what they did and didn't make mistakes.

Now, fast forward to today, pit stops are blazingly fast, with records set in the two to three seconds range. If you've ever seen them work, you'd be amazed. The car's there, you blink your eyes, and it's gone. Now, did they improve efficiency? No, everybody is still lined up in position, with tires and fuel ready, all in a similar manner as pit crews in the 1950s. The efficiency hasn't improved much; they've been doing things without wasted effort for decades. So, how did these pit stops get to be so fast? They focused on effectiveness. They asked the obvious general question:

"How do we make pit stops faster?"

This led to sub-questions:

"How do we get the wheels off?"

"What kind of mechanism do we need to do that?"

"How do we get more fuel into the car faster?"

And they chipped away. By working on their effectiveness, they increased their speed drastically, to the tune of over sixty seconds worth, a 22X savings. That's true effectiveness.

Unfortunately, in the business world, everybody's thinking about being efficient, when they should be looking for effectiveness. To be more effective, ask yourself, *How can I make this process better? How can I make it more impactful so my people don't need to do as much work, or so there are fewer errors in their output?* I've consulted for many companies, and I've witnessed many processes that were efficiently used to accomplish the wrong things. You must get out of this trap and start asking yourself, *Am I doing the **right** things? Could I do them better?* Figure out what to do first, then figure out how to do it efficiently.

Let's demonstrate what this might look like from a technology perspective. Let's say that you own a company that rebuilds roads. Part of your general process includes busting up the asphalt, then loading it onto trucks, then driving it back to your plant. There, machines grind up the asphalt and process it with new sand and aggregate before adding new tar back in. When this is done, the asphalt is dumped back into the trucks and driven back out to the construction site to pave new roads.

Now, if your primary concern is efficiency, you're going to be concerned about getting the trucks from the job site to the plant as efficiently as possible. "Here's our best route to minimize travel time." Yes, that will make the process more *efficient*. But the real question is, "How can we make this more *effective*?" What if you invested in new technology that was portable, allowing you to essentially bring the processing plant to the job site? (You might have even seen this on highway construction sites. Occasionally, you'll find large piles of busted-up concrete or asphalt off to the side of the road, near conveyor belts and a small processing plant.) Now, with your new "mobile mini-plant," you've increased efficiency.

These are the kinds of questions you need to ask whenever you're working with your internal processes:

- *Do I have the right end goal?*
- *How do I measure how long steps are taking to complete?*
- *Do I have any extra steps to finish?*

- *Is there a way to make any of these steps more effective?*
- *How do I measure improvement?*

Companies are riddled with efficient but poor processes. If you want to ride the tsunami, optimize or get rid of those processes. I can assure you, if you have been in business for a number of years, you have processes that could use a bit more effectiveness.

SECURITY

Security is mandatory in any business today. As the world becomes more and more technologically advanced, more and more private data is stored in systems that are vulnerable to attack. Since the proliferation of the internet, we've seen numerous security breaches—at Home Depot, TJ Maxx, Target, and Equifax, to name a few. (Really, the list goes on and on.) When these breaches happen, individuals' identities are put up for sale on the dark web, where criminals use them to secure financing and credit cards, and commit other offenses. The bad news? This trend isn't slowing.

In no way am I advocating for removal of any security processes when it comes to technology. However, I will say that your strategy for approval and review of technology needs to be optimized. You must ensure that security protocols are not imposing undue delays in technology deployment. If you're the owner or CEO, it's your job to ensure that security is implemented and that the people you put in place to manage it don't overburden your company's innovation. If you're a leader in charge of technology, you need to keep this frame of reference to help steer your decisions: it's your job to drive the company forward regarding technology, while at the same time balancing true risk.

As discussed previously, you can't eliminate risk, so remember that as you read through this section.

There are a lot of companies that aren't going to use a lot of the tsunami factors like CNC machines, because they don't produce physical things. We are in the knowledge economy, so many companies produce products that are intangible. Services, consulting, training, and software are just a few of those products. These companies are significantly impacted by cloud computing as a tsunami factor. Thankfully when it comes to security, the advent of the cloud and what is referred to as infrastructure as code (IaC) has made security review easier. IaC is where you deploy infrastructure by writing code versus pointing and clicking or physically racking servers. It's significantly more efficient for deployment. And, if done correctly, it also makes security reviews substantially more efficient.

When you write IaC that can be reviewed by the security department in totality before deployment, they can go through it with a fine-tooth comb and verify that everything's secure. Then when your developers change said code, security doesn't need to do a full review. They can look at the delta (the difference between the two pieces of code) and review only those to ensure that you didn't open up any potential security risks.

Unfortunately, today I still see a very cumbersome and antiquated process in which there are full or multiple reviews for every new deployment. I've even seen situations where security has to program the code that implements controls, which is a horrible practice, because it violates two-party rules. When it comes to security implementation, one person should implement it, and a second person should check and sign off. When you have security doing both of those actions, you lose that safety net.

Security processes are a great place to start looking at for effectiveness. You want to ensure that you aren't putting undue burden or strain on your deployment processes or your TTM. This goes for in-house deployments of software you've developed, and packaged software you've purchased and deployed into the cloud. The rules of engagement are the same.

While reviewing security processes, ensure that you're not trying to "boil the ocean" by eliminating all risk. Again, that's impossible.

Unfortunately, time and time again, I see companies struggle in the security processes area. For instance, one company I worked with had a forty-five-day security review process for implementing firewall rules. I requested a firewall rule be deployed (traffic allowance for software I was deploying), and I waited the forty-five days. Guess what? After reviewing my request and "approving" it, they made a mistake. They only partially implemented my firewall rule, allowing only part of the traffic I requested. It got worse. Common sense would suggest that if they made a mistake, they would go in and correct the error quickly. Unfortunately, the company had lost common sense in its security review process. Since they made a mistake in their implementation, the repair for my firewall rule had to go through another forty-five-day review process. In total, ninety days passed before I could even get my software to work.

I was a consultant. So, I got paid for ninety days to do essentially nothing while I waited for the security department. My wallet was fat and happy, but it cost the company thousands of dollars and months of time, all of which could have been avoided had there been a process without stupidity as a central component.

You don't want to be that company.

FAIL FASTER

To future-proof your organization, learn to fail faster. We're talking about making your entire company agile, not just your software development process. Start using the catchphrase "fail fast" everywhere. It's not just something you should use in technology—the concept applies to your entire business.

For instance, is that marketing campaign not working well? *Fail fast.* Ditch the software package that you thought would be a godsend but turned out to be a nightmare.

Many companies get stuck with tools, software, methods, etc., that aren't working, because they've built time-consuming and rigorous processes around implementing new items. So, once someone gets through all of the hoops and red tape so that they can deploy their new initiative, they don't want to backtrack, even when things aren't working out. The result? Your company continues to push "forward" with a clearly flawed initiative.

If something isn't working or isn't giving the value that you thought it would, get rid of it. Stop where you are and start over. That sounds simple, but in most companies, managers feel the need to save face. They are tired of getting in trouble for canceling something if it doesn't work right. So, they keep going.

Here's my advice: Instead of reprimanding people for trying something and failing, celebrate independence and courage. Recognizing a mistake or when it's time to change directions is a good thing, and it shouldn't be frowned upon. Eliminate obstacles to technological advancement, and that includes making it easy to take a step back and start over.

Startups are very agile across the board. They are running with a limited financial runway and need to succeed as fast as possible. If something isn't working, they stop and get rid of it because they can't afford to waste time.

But the bigger the organization, the more waste there is. Ironically, the waste comes from trying to be efficient. In the name of efficiency, they create more step-by-step processes that supposedly streamline and systemize. Unfortunately, processes that don't make much sense, ones that inhibit the ability to fail fast, often creep into the mix, creating a convoluted scenario. For instance, consider my earlier example from IT security, where the company took forty-five days to create a simple firewall rule. The security vetting process wasn't the only obstacle. There was also the project's approval process. Oh, and then I had to contend with the *software's* approval process. By the time I got the software online and found out that there was a bug and it wouldn't even meet the needs, and it had to be scrapped, months had passed, and dozens (if not hundreds) of man-hours were committed. This situation highlights why companies get stuck trying to make things work. Who wants to go to their boss and tell them to abandon tens of thousands of dollars of effort?

Across the board, your process should allow for an eject button. Importantly, you can't only consider the processes for getting projects on purchases approved. You must also look at the origination processes. With the tsunami, you can't afford to be pushed forward on items that aren't going to net you efficient forward momentum. Try hard, fail fast, abort quickly.

There are two areas where you *do* want to have thorough processes in your company: training, and change management. We'll talk about training first.

TRAINING PROCESSES

In many companies, training is something that you get when you're first hired. Training typically wanes and takes a back seat to productivity after that. With the speed at which technology moves forward, do you think such a one-and-done approach to training is sustainable? When the tsunami makes landfall, do you believe that your organization will last long if you aren't investing in your people's skills? There are two significant factors at play when it comes to training.

First, you need your people to be at the forefront of technology. Continued education will ensure that your people are ready to adopt new technologies. Also, encouraging continued education also enforces many of the key mindsets we discussed in Chapter 4, such as ABC: always be creative.

Secondly, you will increase retention by investing in employees' education and increasing their capabilities. In the future, employee retention will become even more complicated than it is today. With technology moving so quickly, the best and brightest employees want to keep pace. If you're not investing in their progress, they will leave in record time.

When appropriately implemented, here's a strategy that I've seen successfully utilized: a weekly allocation of educational hours. A consulting firm I worked for implemented this, and they saw a substantial improvement in staff skill sets. However, a word of caution—these hours were sometimes abused because there weren't tight controls around what people were "studying," with benchmarks and end goals. It was pretty much a free for all; while many were learning and improving their skills in practical ways, many other employees weren't actually learning anything, were learning absurd things like needlepoint, or were taking forever to learn new skills. But, if you implement educational hours with guardrails and controls, ensuring there are clear expectations about what type of training people are taking and how long it should take, your company will be highly successful.

Unfortunately, no training process will help if you don't have a solid plan for driving change in your organization. So, we'll now cover change management.

CHANGE MANAGEMENT

Throughout the years, I have run into several companies that have implemented fantastic agility in their technology departments, only to have the initiatives stifled by poor change management processes.

Most change management processes treat every new initiative as high risk. But when it comes to modern technology, the opposite is usually true. Modern technology creates more two-way doors than one-way doors, giving you options to revert and roll back to a known working state easier. The ability to take images of servers before major software updates, which allows a rollback point, is one example. Another example of a two-way door is the ability to do a canary deployment. Like the canaries of the coal mines from years past who detected toxic levels of gasses early enough to save people's lives, the canary software deployment saves deployments from affecting your whole customer base.

Canary is a deployment method that enables you to shift a portion of your traffic over to a new version. This traffic division allows you to test a new program with a limited release so that not all of your customers are affected. If there's something wrong, this method allows you to turn off the new version and revert.

Another safety mechanism is a blue-green deployment, where everything is switched over at once, but the old version is still left running, so you can revert to it if there's a problem.

These deployment methods allow companies to take action with significantly less fear of interrupting services. Both canary and blue-green methods mean you no longer need to do upgrades without a point of no return, like in years past. Again, you don't want to stand still; the water is always receding, and you must take action.

These are just a sampling of the risk mitigation techniques available; there are significantly more across the technology spectrum. There are plenty of options that allow you to mitigate risk. Besides, most changes are low-risk anyway, and reviewers should be willing to approve immediately. Only changes that have a point of no return should need to go through a more formal review process. Changes like firewall firmware updates, significant updates to ERP systems, or updates to older legacy systems are all substantial. (But even then, new technologies can mitigate some legacy systems through virtualization and cloud snapshot technology so that you can rapidly roll back changes.)

Overall, you don't want to overburden your change management with mitigating risk. You want to only be formally reviewing high-risk changes. Each technology stack is different, depending on your company and industry. Generally, your goal should be to reduce the burden of change as much as possible, keeping in mind what we learned in Chapter 5 about risk (that most of it is perceived and not solidified in fact).

WRAPPING UP

Process plays a significant part in the overall success of your company. The younger your company is, the fewer formalized processes will exist. So, if you're starting out, it's important that you keep these concepts in mind so that you don't let poor processes—such as bad change management—get in the way. If you're an existing company, you're going to have more work cut out for you as you must fail faster and improve your processes to enhance the overall speed of your organization.

Throughout this book, we've been stacking concepts on top of each other. This chapter built on the discussion in the previous chapter about people. In this chapter, we focused on processes. At this point, you have applied your second layer of wax to your surfboard to ride the tsunami. Hopefully, you're learning how to adapt to stay relevant in the future. In the next and final chapter, we will cover some hidden dangers that may present themselves. You might hit the hidden rocks under the surface if you don't know what you're looking for!

CHAPTER 8

Beware of the Reef

Knowing how to surf isn't the only skill you need to know when riding the technology tsunami. Open water can be dangerous when you don't know what lurks under the surface. You don't want to run into something unexpected while working on streamlining your technology adoption. This chapter will cover items that many companies see as truths and positives but are submerged rocks that will sink your capabilities once you get going.

The first falsehood we are going to cover is the agile fallacy.

THE AGILE FALLACY

The agile fallacy is an interesting concept within the community. There is a fundamental flaw in many people's thinking that equates an agile process with overall business agility. Most of the time, companies think of agile as a destination. You move your software development to agile, and poof! You're future-proof. Nothing could be farther from the truth. Agile is just a methodology for software development and implementation of technology resources. Technically, the agile methodology was specifically made for a software development framework, even though it's also been applied to many other technology implementations in business (such as migrating to the cloud). Often, because businesses need to use the agile method to also remain "agile" in general (able to adapt to a rapidly changing marketplace), the terms get confused. So, for our purposes, I'll use *agile* when I'm writing directly about technology processes, and the word *nimble* to mean that your company must be able to adapt, pivot, and fail quickly in order to respond to the changing technological landscape. *That* is something your business does need. Too many others use the word *agile* for both, confusing people in their company.

As the tech tsunami builds, you're headed in the right direction if you already use the agile software development process. But you can't stop there. Your entire company needs a healthy mindset shift and cultural adjustment to become more nimble. To achieve this, you could add in a couple of the processes we discussed in the last chapter, such as failing faster or improvements to change management. But there are probably one hundred different processes within your organization today that prevent you from getting out of the way of the tsunami. While processes are created to make things more efficient as your company grows, most of the time, they're never revamped to keep up with changes in the business and marketplace.

Agile development is not a cure-all for everything that ails your company. You need to refocus on the macro level and address how your company can become more nimble there. What I'm about to say probably sounds counterintuitive, and it probably won't make much sense when I first say it, but I'll explain:

TECHNOLOGY IS NOT A SOLUTION

I bet you're thinking, *How could technology not be a solution to my problems?* But stay with me.

I have been in technology my entire professional career, since 1995. Before that, I was in technology as a hobby. I've seen many things come and go. One mistake I see repeatedly is that organizations often view technology as a destination, that there isn't somewhere to go after they've deployed that solution, or created that software. Many leaders believe that by implementing one new technology, all their problems will magically be solved. But, companies need to think of technology as a journey rather than a destination.

I've been doing large-scale migrations to AWS, Google, or Azure for Fortune 500 organizations for the last several years. I've worked with some of the biggest names in business, and one thing that I see repeatedly is the use of technology as the end state. I frown upon this kind of thought process, because it's self-limiting.

For instance, I'll hear something like, "We're going to move to the cloud to become more nimble in our company." Okay, moving to the cloud will make you a bit more nimble, but it won't get you all the way there. For that, there's a whole slew of other things within your organization that need to change. Most of the items we've discussed in this book need to be implemented to achieve a truly nimble organization.

So many people fall into this trap. People get lulled into believing marketing hype and sales pitches and throw common sense out the window. For most things, technology is a facilitator, simply a cog in the machine that allows you to make something happen.

Will implementing containerized software speed up your deployment process? By itself, no. Fifty other things need to happen to make faster deployments a reality. Your deployment process is a "process." It's right there in the name. You need to modify that process, which often involves other processes like change management and security review. These things may be made easier due to containerization, but in itself, that isn't the total solution.

Don't fall for the hype and hit those submerged rocks. Keep your view open and see technology as just a piece of your overall solution. You don't want to put your blinders on and think that by implementing some new piece of technology that everything is just going to fall into place. You need to take intentional, strategic action to get things in order. Whenever you implement technology, you should be reviewing the processes attached and the skillsets of the people impacted. By expanding your scope to more than just the technology, you'll ensure that you're meeting the objectives you set. A complete solution includes technology, process, and people.

BEWARE OF SILVER BULLETS

If we look at today's technologies, some are touted as silver bullets. My favorite to pick on is Blockchain. For a while, Blockchain was touted as the be-all and end-all of all technology. Do you have tennis elbow? No problem, just rub on some Blockchain.

The problem is, silver bullets don't exist. There's no way to kill the werewolves lurking around your organization. Blockchain has lost a little steam as the be-all and end-all, but don't worry—AI is on its heels, ready to take its place. I'm not particularly eager to pick on AI, because it *can* solve many problems. But it's also not the be-all and end-all to every problem that's ever existed or will ever exist.

AI is challenging to get up and running. You need to train AI models, and it takes a massive amount of data and labeling to make that happen. So many vendors will talk to you about using AI, but they don't tell you how much effort it takes. Yes, there are out-of-the-box AI solutions, which are great. But if you want anything that moves the needle in your company, it will take a custom AI model, which will require much effort (and cost). The result may be worth the investment, but it's not just something you install and poof! Millions in revenue! Instead, implementing a custom AI model takes time to collect data, time to clean the data, and time to create the models. Can you solve your complex business problem with AI? Possibly. But, can you just snap your fingers and make it happen? Probably not.

The cloud is another technology mistaken for a solution. The cloud is just a tool. It doesn't solve problems, and having worked in the cloud in AWS for years, I can tell you that AWS opens up a can of worms, and there's a shit ton of problems that come out of using it. I'm not saying that the cloud is worse than on-premises, because on-premises has its own can of worms. They're just different cans with different worms. As your environment in the cloud gets more complex, so do your problems. Things like security, audit and compliance, and documentation all get more complicated and harder to manage the greater the environment. There'll be new technology to address those problems, and there'll be problems created from those new technologies, which will require newer technologies, and so forth. As mentioned in Chapter 4, don't be a magazine manager, someone who sees the next shiny object and marketing buzz and simply must have it.

THE WHAT-IF CONUNDRUM

Let's talk about the what-if conundrum. I'm probably not the inventor of the term "what-if conundrum," but for all intents and purposes, let's pretend I am. The conundrum is that, by human nature, when someone says "what if," they're thinking of something negative happening. I don't have any proof, but I'd venture that people go to a negative response 70 percent of the time after saying (or hearing) "what if." What if...we fail? What if...our customers don't like it? Whatever follows "what if" is almost never a good thing.

The whole point of this section is to talk about how to flip the what-if thoughts from fear of action into fear of inaction. Try this the next time you're around your colleagues: whenever somebody says, "What if..." you must stop them immediately and get them to say, "What if we don't..."

When people start saying, "What if we don't," a new, creative mindset will begin to take hold, often illuminating solutions almost automatically. "What if we don't" takes your brain to another level. You must flip the what-if conundrum on its head to change the mindset of your team and your company. If you follow the negative what-if path too far or too long, you will run into the mental blockers we discussed in Chapter 5.

I must admit, I fall into the what-if rut all the time, not so much at work but in my home life. This is an area of growth for me. I try to stop myself before I say, "What if I lose money?" and reframe it to, "What if I don't invest?"

With investment opportunities, *if* I fail to act within the right time-frame, I could miss out on meeting other like-minded investors that lead me to new endeavors.

The moment you start to think *What if I don't...*, your brain evolves, and that's what you want to do within your company. This simple reframing is just one of the ways that you can immediately take action to change the way that people think and operate. We're building surfers that keep their heads up and see their way through problems. You don't want people sitting on the beach, asking, "What if..." and waiting for the perfect wave.

INNOVATION SOURCE

Another possible submerged danger in your organization is not understanding where innovation comes from within your organization. Once organizations decide that, yes, they do need to have a mindset shift around the coming tsunami, they often do one of two things: (1) Create a formal innovation department or (2) create an informal innovation department.

1. THE FORMAL INNOVATION DEPARTMENT

Often, leaders think their organization needs some snazzy new department that creates incredible off-the-wall products and ideas—a think tank full of super bright people whose job is to "innovate"—you know, a skunkworks, an innovation hub. On the contrary, that's exactly what you *don't* need.

Every skunkworks—or innovation hub, or department for innovation—I've ever seen (and I've seen a lot) has failed. There are several reasons why.

First, these skunkworks work in a vacuum. So, they don't understand the actual needs of their organization. Many leaders observe the skunkworks at Northrop Grumman or a similar institution and try to emulate their setup. But the think tanks at those institutions have problems to solve, and they are tasked with solving those specific problems. From experience, I know that most organizations create a think tank whose job is to identify the problems themselves. The issue is, the think tanks frequently *do* come up with problems. Unfortunately, the problems they come up with are often artificial constructs that may not exist in the real world.

I worked for a consulting firm that was in the small and medium business IT space. They had an innovation hub to which they gave a large amount of funding and autonomy. I like their idea in theory—it makes sense to give a think tank money and freedom to do their job. However, the think tank didn't have a specific task to complete. Instead, they just went on their merry way, conjuring up solutions for customers. The team didn't have any salespeople or sales engineers. It was made up mostly of project managers and executives with fancy titles.

The think tank turned out several products in a fairly short amount of time. They were very efficient. Their products came complete with go-to-market strategies, sales presentations, and marketing materials. The only problem was that when the rest of the company went to sell these products, we couldn't. The solutions seemed great, but there was no market for them. No one wanted the think-tank solutions.

If you are going to have a formal innovation department, use them the right way: feed them the most complex problems that you need solved *for your customers,* and give the think tank the autonomy to solve them. Whatever you do, don't let them invent problems.

2. THE INFORMAL INNOVATION DEPARTMENT

The second problem I see around innovation occurs when organizations leave it to a select few people. In this scenario, the company's innovation champions aren't technically a separate team or department. They are just innovation-friendly individuals scattered throughout the firm. This solution is significantly better than the formal innovation department because at least the innovators aren't working within a vacuum. However, this solution still doesn't meet the requirements for the technological future.

Instead, I encourage you to seek out and develop as many people across your organization as possible. Innovation is everyone's job. So, have everyone solve problems and consider and develop new solutions. For starters, encourage others to bring forth issues they see. You would be surprised how many times I've heard people complaining about a problem that no one has ever taken to management. Also, allow people to experiment. Experimentation is crucial, as it encourages the new mindset that you're trying to create. Finally, if someone wants to work on solving a problem for another team or department, please, for the love of all things holy, don't tell them, "That's not your job." Instead, celebrate and encourage their work ethic and excitement.

When you delegate innovation across the institution, you'll be tapping into the collective brainpower of your organization to solve problems. If you're a manager, you can encourage your staff to be innovative. Show them that you're on board with experimentation, and they won't be reprimanded for failure. Southwest Airlines is a good example of a company that does this well. They encourage and empower everyone to solve problems. It doesn't matter if you're an executive or a baggage handler: everyone is considered an innovator. Southwest is the leader in plane turn-around times. They can get a plane unloaded and reloaded faster than any other airline. They've accomplished this by allowing their ground crew to be innovative and experiment with new methods.

LOOK OUTSIDE

There is one other issue I often see: companies tend to only look inside for innovation. But, that's a limiting view. A better perspective would be to look at your *customers'* perspectives, views, and ideas, and then use that information for innovation.

Software companies use their customers' ideas quite well. It's standard in that industry to have an avenue (like a link in the app or website) where customers can submit feature requests. For instance, a significant percentage of the features of and changes to Amazon Web Services stems from client requests. Customer requests drive around 80 percent of AWS feature development, which is quite impressive.

From a mathematical position: there are probably tens of thousands of people who work in AWS, but there are *millions* of external users. Those millions could be the key ingredient to an unstoppable force of innovation. Your customers will almost always have more collective brainpower than your company does internally. Tap into that brainpower, harness that outside innovation, and help further your organization internally to develop new products and processes.

Now, *how* you implement this customer need/desire feedback loop depends on the type of organization you work for. In software, it's relatively easy to have a feedback loop by simply including a feature request link somewhere in your application. But it's a bit more complicated for other industries. The fundamental aspect is creating the cultural shift around customer feedback. Likely, you'll need to coach some staff who have a "not invented here" attitude. (That is, they think they know what's best and have an elitist view on feedback.) If you work through innovation hurdles, you will solidify your safety as you ride the wave.

R&D BUDGET

The lack of a dedicated budget for research and development isn't so much of a submerged rock but rather the cliff's rock face that you can run into. Of all the small and medium companies I've worked for, none of them have had an R&D budget. In my opinion, that's a crime. This lack of funding makes innovation an afterthought. If you aren't planning to innovate, you're planning to fail.

Say you work as an engineer in the IT department at a company that has a custom-built, memory-based database for commodities trading. You'd like to increase its performance with a new memory technology that's available. But, the technology is groundbreaking, and there's not a lot of information available about it. So, it's not mainstream yet. Still, from your perspective, your company should get behind it, as it will likely drastically increase your competitiveness by reducing order fill time.

You know that implementing the memory technology will require some additional computer programming. You go to management and make your case that your company should invest in this new technology. But the conversation doesn't go well, because there's no predetermined line item in the budget for experimentation.

So, you're making the request as a one-off need, and they decline to give you the resources—you get neither no new hardware nor the engineering time required to implement the new technology. Without an R&D budget, a company is unlikely to respond to impromptu technological needs and advancements, and the adage of needing to spend money to make money is 100 percent true. R&D directly correlates to your company's ability to innovate to drive revenue or decrease expenses.

Without a true R&D budget, your company won't have the resources to adapt in tough times. If you start to lose traction in the market due to competitive pressures, you don't want to be stuck in a position where you don't have funding to implement new capabilities. In that situation, your company can quickly get into a death spiral where you're losing revenue to a competitor while being unable to drive your business forward.

But *with* an R&D budget, you're essentially protecting your ability to innovate. Companies typically resist allocating money to R&D, and this resistance compounds when your company is in financial distress. Take, for instance, restaurants during COVID-19. If an owner or management had money set aside for R&D, they were able to make investments in online ordering technology. However, those that didn't have an R&D budget (and many didn't) wound up in too deep of a hole to dig themselves out, and, oftentimes, their businesses failed.

Additionally, many people also see R&D as frivolous, as it doesn't provide immediate value, and without money earmarked by management, most will never even attempt to drive innovation that doesn't yield immediate results. So, without a budget for it, any new innovation project will fail before it even starts.

If you're a manager, you should be earmarking money within your own department for R&D. There is nothing saying that just because the company doesn't fund it, your budget can't. By allocating your budget to R&D, you empower your team to innovate and drive the business forward.

If you do create a dedicated R&D budget, ensure that it cannot be pillaged to cover other business expenses. It must be a sacred fund for the specific function of research.

DON'T PENALIZE FAILURE

The last critical danger that you need to avoid is penalizing failure. I've never been huge on punishing failure for any reason. Punishment demoralizes people, and you don't want to discourage your team from experimenting. A demoralized team doesn't work as hard. They aren't willing to go the extra mile, and they're unwilling to put their ass on the line for an idea. Let's focus on that last point.

When we're talking about technological innovation, you need people who are willing to step up to the plate and take a swing for the fences, to give it all they have with a new technology, concept, or idea. But those are the very experiments that may propel your company into the future. If you work in an environment where failure is punished, you're not going to be successful, because people will eventually stop risking themselves for their innovative ideas.

Once, I worked at a medium-sized financial services company. Overall, it was an OK company to work for. We had a typical atmosphere around failure—failure wasn't desired, but, to be fair, they didn't treat failure as if it were the end of the world, either.

Unfortunately, I got a new boss, Andy, who became the CIO. They changed the attitude around failure into a witch hunt.

I'm not too fond of that mindset. When something fails, it's not a failure of the *person*; it's a failure of the *process*. When you or your team members make mistakes, you need to know what happened, not who did it. After you know what happened, you can figure out why it happened. Then, you can determine if there's a way to fix the problem or process so it doesn't happen again. We're all human. We all make mistakes, and to me, that's acceptable. When people keep making the same mistakes repeatedly, that's not acceptable.

Unfortunately, Andy didn't see things the way I see them. He would specifically seek out ideators to find out "who did it," which was very demoralizing. Soon, no one was willing to stick their necks out anymore for a potentially good idea. Andy had succeeded in creating a culture of fear, which is the worst thing you can have when you want to remain innovative and competitive. If you make your staff fearful of making mistakes, they will do everything in their power to ensure that they never make a mistake. How do you never make a mistake? You do the same, monotonous thing, repeatedly. You follow the beaten deer path in the woods.

I soon left the company. Andy still works there, and based on what I've heard, their innovation has significantly suffered. They are running on out-of-date operating systems, and no new innovations have taken place since my departure.

If you want to keep a culture of innovation, never, ever penalize someone for failure. (Now, if people continue making mistakes because they're not following set processes, that's entirely different.) The culture you *do* want is one where failure is actually encouraged. As I tell my daughters, "If you're not failing, you're not trying hard enough, and you're being complacent." One of my daughters was a straight-A student in middle school. I told her:

> That's great! I'm very happy that you get straight As now. But, when you get to high school, I don't want you to get straight As. I want you to take more challenging courses, to keep going higher and harder until you fail.

With that sort of attitude, two things happen. First, you find your true limits. By pushing, my daughter could have found that she was even better at some subjects than she imagined. Or, maybe she'd find out that Advanced Placement Spanish wasn't really her thing, which she'd never find out if she didn't try. When you find your limits, you have the option to continue by putting in more effort, or to redirect your efforts elsewhere. But if you don't challenge yourself, and you instead take the easy road to keep getting the "As," you'll just float through life, never identifying your full potential.

The same thing is true for your organization. You want each person to push themselves on an individual level. You want people to reach their failure points, to know if they should try harder or focus somewhere else. Secondly, you want your *organization* to act the same way. You want to push the limits of your organization until it can't go any further. If you create a culture of fear, and there is a stigma around failure, you're never going to find your organization's potential. Instead, you're only going to be as good as the weakest link of power within your organization. Instead, drive your organization using these methods and reach new heights.

BEWARE THE ECHO CHAMBER

When companies operate for a long duration, there can be an effect called an echo chamber. An echo chamber is when the conversations that take place and the people involved all support the current path. I've seen this manifest from not wanting to move off of existing tools to not changing processes. Let's say, for instance, that I make a widget and I've been using metal castings to make this widget for the last decade. If I were to say to my team of employees that have all been there for the last decade if casting is the best method, what do you think I will have echoed back? There is a high probability that I'll get a resounding "Yes, casting is the best method." This echo occurs because everyone on my team knows the technology and hasn't been outside of the team to gain any new experiences.

As you can probably see, when it comes to moving your company forward, these echo chambers can be very dangerous. Most companies see a lot of employee rotation, with the tenure of tech employees lasting only a couple of years. This churn of employees keeps the echo chamber effect low. But some industries, like community banking, don't have this churn. It's not uncommon to see people working at the same company for a decade or two in this industry. The lack of new innovative ideas being injected into your company from the outside amplifies the echo chamber effect.

If your company or industry has very long tenure, you need to take a proactive approach to reducing the echo chamber effect. You can accomplish this through job rotation and training to help expand the experiences of your employees. Encouraging their growth will not only improve your technology stance but also your culture and employee fulfillment.

WRAPPING UP

We've come to the end of our journey. We talked about what the tsunami is, and when it originated. Once we had that covered, we talked about the tsunami factors and why now is such a critical point in technological history. Then, we started discussing what you can do about it; that's when we covered the heartache that other companies have gone through. Their stories act as lessons so you can start to visualize risk accurately within your own company. We talked about how to surf, finding a new mindset, and the necessary cultural and process shifts to remain competitive. Finally, we covered the hidden dangers that you need to avoid.

Technology is so diverse that there are infinite possibilities. There isn't enough space to put in all of the potential dangers lurking under the water's surface. This chapter intended to touch on the items that most everyone will run into during their surfing journey.

You now know what the technology tsunami is and how it will impact you and your organization. Furthermore, you understand how you need to adjust your mindset and get past mental blockers. Finally, you know how to wax your surfboard to ride the wave and take your company into the future. Now, go ride.

Conclusion

The sun is bright and sweltering hot. You have your wetsuit on, and you're lying on top of your board as you paddle out into the open water. The water is growing higher and higher. You paddle frantically to catch the peak of the wave. But this isn't any wave; it's a tsunami. It's not going to be easy to catch. There will be an enormous amount of work ahead of you.

But, guess what? Now, you know *how* to catch it. You won't be floundering about trying to find your way. Your board is waxed, your skills are strong, and you have the mental awareness to deal with the fear and uncertainty. You've got this.

I formulated this book to give you a firm understanding of the types of issues that companies will face in the future—showing you our innovative past, present, and future by demonstrating the tsunami effect of innovation and the massive push that technology will have in our near future. Let's take a few moments to recap and bring the entire picture together.

Chapter 1 covered what the tsunami is and where it started. I demonstrated this by showing the compounding effect of technology using the blacksmith and the water bottle examples. From these examples, you learned that there is a massive difference between the rate of innovation in ancient history and today. What took millions of years now takes mere years or months. Once you understood the present rate of ever-accelerating change, you understood the importance of getting your ass in gear now.

Understanding the tsunami brought us to Chapter 2. This chapter highlighted all of the modern technologies and advancements that have set humanity up for critical mass, which I call the tsunami factors. Now is the time to take action to redevelop how your company adopts technology. Action will be vital to your company's survival, and success.

But we still needed a baseline of what kind of disruption the tsunami will bring. In Chapter 3, I addressed this by highlighting some company failures of the recent past, such as Sears, Kodak, and Blockbuster, which were slow to react to technology and didn't understand the true impact of what was happening in the marketplaces around them. They may have seen a swell in the ocean, but they weren't expecting it to wash them away. Armed with their stories, you now know that even seemingly benign technology may have a massive effect on your business. To demonstrate the impact of technology, I compared and contrasted the heartache of getting left behind versus the heartburn of doing what must be done to remain viable.

Those chapters concluded the first part of the book. They set the stage, and the danger. To give you a fighting chance, part two of the book focused on critical items that you could address in your organization to keep things afloat: In Chapter 4, we covered the mental preparedness you need to succeed. Technology will be coming at you like water out of a firehose. Outdated, even once-successful thought processes aren't going to cut it in the future. So, we focused on your new way of thinking—like ABC (always be creating) and how to think like a child. It's only through a wider lens of the world that you will truly see the hidden technology opportunities for your business.

But unfortunately, even with the right mindset, not everything is rainbows and unicorns. A significant number of mental blockers will also come into play as you're trying to ride the tsunami. That's why, in Chapter 5, I covered some of the major mental blockers. We discussed risk at length—about how it affects us, our employees, and our companies. We learned that risk isn't something to be eliminated but accepted. We also covered other mental blockers such as compensation turmoil and external pressures. These items will need to be addressed to move your organization forward.

With an excellent mental picture, we covered what changes will need to be enacted to properly prep your organization. Chapter 6 dealt with the cultural changes that must take place to surf the tsunami. You want your people to be soft and supple like a well-oiled boot, so that you can harness the power of creativity and continual evolution.

In Chapter 7, we covered process changes. Even though changing processes isn't always fun, it's significantly more straightforward to change than culture. A key point in this chapter was the difference between efficiency and effectiveness. Process effectiveness reigns supreme in driving your organization further. We also touched on legacy processes, such as procurement, that slow down your technology adoption.

Finally, in Chapter 8, we discussed some submerged rocks or potential pitfalls common in the technology adoption process that pose unseen problems. Having the knowledge to look ahead and under the surface allows you to see these submerged rocks. Items such as the lack of R&D budgets as well as supposed silver bullets are significant threats to your ability to surf the technology wave.

We have now reached the end of our journey together. It's now up to you to move your company forward and keep riding the wave. Although this book isn't a paint-by-numbers solution for your company's problems, you should now have enough background to make the necessary changes. If you need more help or resources, check out my website, aaronalfini.com, for additional information and training. Surf's up!

Acknowledgements

A book is not a one-person journey. Although my name is on the cover, this book wouldn't be possible without the support of at least a dozen people. The first person that I'd like to thank is my wife for putting up with me and supporting me in all of my endeavors. Without her, I wouldn't have finished this project. I, of course, also have to mention my daughters, Natalie and Hailey, for giving up some dad-time so that I could write, edit, and obsess about my book. Whether they know it or not, my family gives me the drive to continue moving forward, even when it's painful and I want to give up.

A special thanks goes out to all of my beta readers—John Kolosci, Fiona Greely, Jeff Melbostad, Jeff Hoffman, and Randy Strick—for providing feedback to clean up the initial version. Finally, thanks to my editor Paul Fair and all of the other awesome people at Scribe Media.

About the Author

AARON ALFINI has been adopting technology since the second grade, when he learned to program on a Commodore 64 with a black and white TV. He doesn't see himself as a technologist, but as a creative, with technology as his canvas.

Aaron has worked for startups and Fortune 100 companies alike. He's worked on some of the largest cloud migrations in the world, with clients including AWS, Equifax, Getty Images, Discovery Communications, Met Office, The London Underground, Mattel, and US Bank.

A lifelong learner, Aaron has a bachelor's in business administration, a master's in cyber security, an executive certificate in strategy and innovation from MIT, and a certificate in executive leadership from Cornell.

Aaron lives in Chicago with his wife and two teenage daughters. He is a tinkerer who enjoys countless hobbies, such as restoring cars, being a pilot, breeding ball pythons, sailing yachts, collecting WWII firearms, woodworking, and smoking cigars while sipping on one of the countless bourbons in his collection.

www.ingramcontent.com/pod-product-compliance
Lightning Source LLC
Chambersburg PA
CBHW031855200326
41597CB00012B/416